Lecture Notes in Economics and Mathematical Systems

For information about Vols. 1–128, please contact your bookseller or Springer-Verlag

Vol. 129: H.-J. Lüthi, Komplementaritäts- und Fixpunktalgorithmen in der mathematischen Programmierung. Spieltheorie und Ökonomie. VII, 145 Seiten. 1976.

Vol. 130: Multiple Criteria Decision Making, Jouy-en-Josas, France. Proceedings 1975. Edited by H. Thiriez and S. Zionts. VI, 409 pages. 1976.

Vol. 131: Mathematical Systems Theory. Proceedings 1975. Edited by G. Marchesini and S. K. Mitter. X, 408 pages. 1976.

Vol. 132: U. H. Funke, Mathematical Models in Marketing. A Collection of Abstracts. XX, 514 pages. 1976.

Vol. 133: Warsaw Fall Seminars in Mathematical Economics 1975. Edited by M. W. Loś, J. Loś, and A. Wieczorek. V. 159 pages. 1976.

Vol. 134: Computing Methods in Applied Sciences and Engineering. Proceedings 1975. VIII, 390 pages. 1976.

Vol. 135: H. Haga, A Disequilibrium – Equilibrium Model with Money and Bonds. A Keynesian – Walrasian Synthesis. VI, 119 pages. 1976.

Vol. 136: E. Kofler und G. Menges, Entscheidungen bei unvollständiger Information. XII, 357 Seiten. 1976.

Vol. 137: R. Wets, Grundlagen Konvexer Optimierung. VI, 146 Seiten. 1976.

Vol. 138: K. Okuguchi, Expectations and Stability in Oligopoly Models. VI, 103 pages. 1976.

Vol. 139: Production Theory and Its Applications. Proceedings. Edited by H. Albach and G. Bergendahl. VIII, 193 pages. 1977.

Vol. 140: W. Eichhorn and J. Voeller, Theory of the Price Index. Fisher's Test Approach and Generalizations. VII, 95 pages. 1976.

Vol. 141: Mathematical Economics and Game Theory. Essays in Honor of Oskar Morgenstern. Edited by R. Henn and O. Moeschlin. XIV, 703 pages. 1977.

Vol. 142: J. S. Lane, On Optimal Population Paths. V, 123 pages. 1977.

Vol. 143: B. Näslund, An Analysis of Economic Size Distributions. XV, 100 pages. 1977.

Vol. 144: Convex Analysis and Its Applications. Proceedings 1976. Edited by A. Auslender. VI, 219 pages. 1977.

Vol. 145: J. Rosenmüller, Extreme Games and Their Solutions. IV, 126 pages. 1977.

Vol. 146: In Search of Economic Indicators. Edited by W. H. Strigel. XVI, 198 pages. 1977.

Vol. 147: Resource Allocation and Division of Space. Proceedings. Edited by T. Fujii and R. Sato. VIII, 184 pages. 1977.

Vol. 148: C. E. Mandl, Simulationstechnik und Simulationsmodelle in den Sozial- und Wirtschaftswissenschaften. IX, 173 Seiten. 1977.

Vol. 149: Stationäre und schrumpfende Bevölkerungen: Demographisches Null- und Negativwachstum in Österreich. Herausgegeben von G. Feichtinger. VI, 262 Seiten. 1977.

Vol. 150: Bauer et al., Supercritical Wing Sections III. VI, 179 pages. 1977.

Vol. 151: C. A. Schneeweiß, Inventory-Production Theory. VI, 116 pages. 1977.

Vol. 152: Kirsch et al., Notwendige Optimalitätsbedingungen und ihre Anwendung. VI, 157 Seiten. 1978.

Vol. 153: Kombinatorische Entscheidungsprobleme: Methoden und Anwendungen. Herausgegeben von T. M. Liebling und M. Rössler. VIII, 206 Seiten. 1978.

Vol. 154: Problems and Instruments of Business Cycle Analysis. Proceedings 1977. Edited by W. H. Strigel. VI, 442 pages. 1978.

Vol. 155: Multiple Criteria Problem Solving. Proceedings 1977. Edited by S. Zionts. VIII, 567 pages. 1978.

Vol. 156: B. Näslund and B. Sellstedt, Neo-Ricardian Theory. With Applications to Some Current Economic Problems. VI, 165 pages. 1978.

Vol. 157: Optimization and Operations Research. Proceedings 1977. Edited by R. Henn, B. Korte, and W. Oettli. VI, 270 pages. 1978.

Vol. 158: L. J. Cherene, Set Valued Dynamical Systems and Economic Flow. VIII, 83 pages. 1978.

Vol. 159: Some Aspects of the Foundations of General Equilibrium Theory: The Posthumous Papers of Peter J. Kalman. Edited by J. Green. VI, 167 pages. 1978.

Vol. 160: Integer Programming and Related Areas. A Classified Bibliography. Edited by D. Hausmann. XIV, 314 pages. 1978.

Vol. 161: M. J. Beckmann, Rank in Organizations. VIII, 164 pages. 1978.

Vol. 162: Recent Developments in Variable Structure Systems, Economics and Biology. Proceedings 1977. Edited by R. R. Mohler and A. Ruberti. VI, 326 pages. 1978.

Vol. 163: G. Fandel, Optimale Entscheidungen in Organisationen. VI, 143 Seiten. 1979.

Vol. 164: C. L. Hwang and A. S. M. Masud, Multiple Objective Decision Making – Methods and Applications. A State-of-the-Art Survey. XII, 351 pages. 1979.

Vol. 165: A. Maravall, Identification in Dynamic Shock-Error Models. VIII, 158 pages. 1979.

Vol. 166: R. Cuninghame-Green, Minimax Algebra. XI, 258 pages. 1979.

Vol. 167: M. Faber, Introduction to Modern Austrian Capital Theory. X, 196 pages. 1979.

Vol. 168: Convex Analysis and Mathematical Economics. Proceedings 1978. Edited by J. Kriens. V, 136 pages. 1979.

Vol. 169: A. Rapoport et al., Coalition Formation by Sophisticated Players. VII, 170 pages. 1979.

Vol. 170: A. E. Roth, Axiomatic Models of Bargaining. V, 121 pages. 1979.

Vol. 171: G. F. Newell, Approximate Behavior of Tandem Queues. XI, 410 pages. 1979.

Vol. 172: K. Neumann and U. Steinhardt, GERT Networks and the Time-Oriented Evaluation of Projects. 268 pages. 1979.

Vol. 173: S. Erlander, Optimal Spatial Interaction and the Gravity Model. VII, 107 pages. 1980.

Vol. 174: Extremal Methods and Systems Analysis. Edited by A. V. Fiacco and K. O. Kortanek. XI, 545 pages. 1980.

Vol. 175: S. K. Srinivasan and R. Subramanian, Probabilistic Analysis of Redundant Systems. VII, 356 pages. 1980.

Vol. 176: R. Färe, Laws of Diminishing Returns. VIII, 97 pages. 1980.

Vol. 177: Multiple Criteria Decision Making-Theory and Application. Proceedings, 1979. Edited by G. Fandel and T. Gal. XVI, 570 pages. 1980.

Vol. 178: M. N. Bhattacharyya, Comparison of Box-Jenkins and Bonn Monetary Model Prediction Performance. VII, 146 pages. 1980.

Vol. 179: Recent Results in Stochastic Programming. Proceedings, 1979. Edited by P. Kall and A. Prékopa. IX, 237 pages. 1980.

Vol. 180: J. F. Brotchie, J. W. Dickey and R. Sharpe, TOPAZ – General Planning Technique and its Applications at the Regional, Urban, and Facility Planning Levels. VII, 356 pages. 1980.

Vol. 181: H. D. Sherali and C. M. Shetty, Optimization with Disjunctive Constraints. VIII, 156 pages. 1980.

Vol. 182: J. Wolters, Stochastic Dynamic Properties of Linear Econometric Models. VIII, 154 pages. 1980.

Vol. 183: K. Schittkowski, Nonlinear Programming Codes. VIII, 242 pages. 1980.

continuation on page 185

Lecture Notes in Economics and Mathematical Systems

Managing Editors: M. Beckmann and W. Krelle

269

Anders Borglin
Hans Keiding

Optimality in Infinite Horizon Economies

Springer-Verlag Berlin Heidelberg GmbH

Managing Editors

Prof. Dr. M. Beckmann
Brown University
Providence, RI 02912, USA

Prof. Dr. W. Krelle
Institut für Gesellschafts- und Wirtschaftswissenschaften
der Universität Bonn
Adenauerallee 24–42, D-5300 Bonn, FRG

Authors

Prof. Anders Borglin
Department of Economics
University of Gothenburg
Box 3091, Gothenburg, S-400 10 Sweden

Prof. Hans Keiding
Institute of Economics
University of Copenhagen
Studiestraede 6, DK-1455 Copenhagen K, Denmark

ISBN 978-3-540-16475-3 ISBN 978-3-662-02478-2 (eBook)
DOI 10.1007/978-3-662-02478-2

Originally published by Springer-Verlag Berlin Heidelberg New York Tokyo in 1986

TABLE OF CONTENTS

PART II

ACKNOWLEDGEMENT

Bodil Olai Hansen has read part of the manuscript and thereby saved us from some mistakes and misprints. We want to use this opportunity to thank her as well as Erling Petersson, who has helped us conquer the mysteries of the word processor, and Agnetta Kretz, who has drawn the diagrams.

The Swedish Social Science Foundation and the Royal Swedish Academy of Science has provided financial support for Anders Borglin.

INTRODUCTION

Modern welfare economics as it is known today to economists took its final shape with the emergence of the Arrow-Debreu model. The classical conjectures about the beneficient workings of markets together with the converse statement, that optimal (in the sense of Pareto) allocations may be sustained by prices and markets, has laid a firm foundation for further research in welfare economics. But more than that, it has inspired researchers to take up entirely new topics, notably by closer considerations of situations where the assumptions of the original model may seem overly restrictive.

One of these new directions has been connected with generalizing the model so that it takes into account the possibility of infinitely many commodities. On the face of it, the idea of an infinity of commodities may seem a mathematical fancy having no "real" counterpart in economic life. This is not so, however. Quite to the contrary, infinity enters in a very natural way when it is taken into account that economic transactions take place over time.

In the Arrow-Debreu formalism, time may be incorporated into the model in a very simple way using dated commodities. Thus two commodities are considered as being different if they are to be delivered at different points of time. This opens up new interpretations of the model as describing economies over time, but obviously it presupposes that only a finite set of time points (or dates) are of importance, or in economic terms, that the agents of the economy have a common finite horizon. Without this assumption we cannot use the standard theory with dated commodities.

Unfortunately, this assumption is not readily acceptable. It may be reasonable to assume a finite horizon for any single consumer, but it is strictly less so when we speak of a firm, for although we have no accounts of eternal firms we cannot on the other hand put a common limit to the lifespans of all firms (unless we let it be of astronomical order of magnitude). Even more difficult would it be to put a finite horizon on an economy: every single agent may have a limited(and very short for that sake) lifespan, but other agents (consumers and firms) take over, and what is the really crucial point, economic transactions take this into account since a great number of contracts presuppose the future actions of yet unborn generations (pension schemes is the standard example, government borrowing is another one of considerable international significance, but it goes for very many financial agreements, the famous British consols of the eighteenth century being perhaps the most clearcut illustration).

Having realized the importance of models with no fixed horizon, the logical next step is to construct models with an infinite number of commodities, and actually this development followed quite quickly after the emergence of the Arrow-Debreu model. Lately one particular type of such models, namely models with overlapping generations have become increasingly popular as a tool for investigating various phenomena. The distinguishing feature of such models, as compared to more general models with infinite commodity space, is that the possible actions of each consumer or producer is confined to a finite dimensional subspace. Since there is in each generation or, in more general models, on each market only a finite number of agents it is possible to apply many of the techniques developed for economies with a finite number of agents. It appears to be a fruitful view to consider this type of economy as a family of finite economies having the possibility of trade between each other.

The possibility of applying many of the techniques developed for finite economies provides an explanation for the abundance of applications of overlapping generation economies. Among these are applications to monetary theory, finance, rational expectations, equity and justice, bequests, national debt, foreign debt and the theory of exhaustible resources. Although long, this list is by no means exhaustive.

As is evident from its title the main topic of this book is the welfare economics of overlapping generations models. For such a model Pareto optimality and production efficiency may be defined verbatim as for an economy with a finite commodity space. Nevertheless with an infinite commodity space the standard results mentioned at the outset, in particular, the First Theorem of Welfare Economics, turn out no longer to be true. A simple example of this failure is the following one, drawn from capital theory. At date 1 there is a (marvellous) cake whose quality is unchanged no matter how long it is stored. Thus at each date a portion of it may be consumed and the rest stored. Suppose that the cake-price was increasing over time (we do not need to wonder why it should be so at this place; this would obtain if the consumers of each generation were suitably patient so that they preferred consumption tomorrow rather than today). Then it would be a profit maximizing action for the storage-producing firm to keep the cake intact from each and every date to the next. On the other hand it is easy to find another program which will make every consumer better off: cut a fixed proportion of the remainder of the cake at each date and divide it among the consumers. If consumers prefer some cake to no cake (which does not seem too unreasonable) this is an improvement for everyone, showing that the profit maximizing action leads to inoptimality.

On a general level it turns out that the concepts of Pareto optimality and production efficiency in economies with a finite number of commodities correspond to weaker notions of optimality and efficiency in overlapping generations economies, namely what is known as <u>weak Pareto optimality</u> and <u>weak production efficiency</u>. Thus a feasible allocation, for which there is a price system supporting each agents action, has the property that there is no alternative feasible allocation, differing from the given one only for <u>finitely</u> many agents , which is considered at least as satisfying as the original allocation by all consumers and more satisfying by some. Conversely, for a feasible allocation having this property there is a price system supporting the individual actions.

This leads us to the main topic of this book, the first part of which treats generation models and some related topics in the case of a single good at each date, while Part II is devoted to a generalization and extension of these results to more general models. In Chapter 1 we give a review of some of the earlier contributions and illustrate, by means of examples, the problems at hand. As it will be seen from Chapter 1, there are important common features of the production model and the consumption model. In Chapter 2 we exploit these similarities to define reduced models. These capture the substitution possibilities between adjacent dates. Some examples of reduced models arising from related models are given as well. The particular structure of generations economies makes it possible to define a composition on the sets making up a reduced model. This is done in Chapter 3, for one good models and in Chapter 8 for many-goods models. A general criterion for efficiency of reduced models is then developed; in Chapter 3 for one-good models, and in Chapter 8 is given a corresponding exposition for many-goods models.

A good many contributions have given various parametric criteria for efficiency or Pareto optimality of weakly efficient (weakly Pareto optimal) allocations. These criteria arise from consideration of parameters describing the economy; typically supporting prices or some measure of curvature. In Chapter 3 we show how such parametric efficiency criteria are related to the general efficiency criterion by using the semi-group structure inherent in reduced models and relating it to different semi-group structures on the real line.

Some of the results in Chapter 3 are derived under the assumption that the reduced model has a support with all prices equal to one. This may always be achieved by a suitable choice of the units of measurement for the goods and, as the results of Chapter 4 show, the criteria derived generalize to models with arbitrary support. In Chapter 4 is also shown that parametric efficiency criteria are closely related to approximations of a given model with sets belonging to some small family indexed by the associated measure of curvature. The main part of the chapter is concluded with a discussion of different measures of curvature and their related families of approximation. Finally, some further efficiency criteria for the discounted utilty model and the pure birth process, which were studied in Chapter 2, are given.

The understanding of the relation between the general efficiency criterion and parametric efficiency criteria as developed in Chapter 3 and 4 paves the way for the results given in Chapter 5, for the one-good case, and Chapter 9, for the many-goods case. In Chapter 5 we introduce axioms for a measure of curvature and show how these axioms determine uniquely a measure of curvature and a corresponding family of approximations. A slight reformulation and extension in Chapter 9 makes it clear that most of the reasoning carries over to many-goods models.

For the one-good models of Part I the market structure is fixed. In Chapter 6 we study, for the many-goods case, possible market structures and economies defined on these. The market structure is to some extent arbitrary and we make use of this in order to define a canonical market structure which is, loosely speaking, the type of market structure occurring in a generation model. The indeterminacy of the market structure makes it desirable to introduce mappings, called morphisms, between market struktures and models defined on these. Through the use of such morphisms it is possible to abstract from irrelevant details of the models and stress their common aspects. In Chapter 7-9, we take advantage of this possibility. In fact morphisms were applied already in Chapter 4 in order to change the units of measurement. For the many-goods case morphisms can, when suitably defined, handle both a change in the units of measurement and a variation in the market structure.

In Chapter 7 we carry out a program for many-goods parallelling the one for one-good models in Chapter 2. Although much of the reasoning of Chapter 2 applies certain care must be taken as the market structure as no longer fixed. Similarly the contents of Chapters 8 and 9 parallel that of Chapter 3-5 of Part I.

Throughout the book we employ standard mathematical notation. The set of natural numbers is denoted by N, the set of real numbers by R. Further, R^k is the k-dimensional Euclidean space, R^k_+ its non-negative orthant, i.e $R^k_+=\{x \varepsilon R^k \mid x_h \geq 0,\ h=1,\dots k\}$. The set-theoretical membership is denoted by ε (for typographical reasons)and for B a set in a topological space, intB denotes the interior and clB the closure of B. The sign * after the number of a section indicates that the material will not be used in the sequel. Finally ● marks the end of a definition, a proof, or an example.

PART I

ONE-GOOD MODELS

CHAPTER 1: ONE-GOOD PRODUCTION AND CONSUMPTION MODELS

The present chapter is an introductory one. We describe the infinite horizon production and consumption models which provide the background for efficiency considerations without fixed horizon. The reader already acquainted with these models may proceed directly to Chapter 2.

The models of this chapter - indeed in all of part I - are simple in several respects; there is only one good at each date, and each economic agent, consumer or producer, is concerned only with actions at two consecutive dates. Also, in this chapter there are at most two agents acting at each date.

The concepts of efficiency and optimality, which are central for this book, are introduced and the fundamental problem in infinite horizon models - the non-correspondence of efficiency and the property of being supported by a price system - is elucidated.

Clearly, the models described in this chapter can be, and have been used for other purposes than ours - for investigations of the burdens of different generations, the theory of interest and the role of money- to mention only some. However, to some extent such problems do involve efficiency/optimality considerations, and a solution to this one problem may yield considerable insight into the other problems.

1.1. THE ONE-GOOD PRODUCTION MODEL

The one-good production model has at each date a single good which may be consumed or invested as (shortlived) capital to give production at the next date. The well-known growth model of Swan [1956] and Phelps [1965] involves both labour and capital. Our first step is to see how this model in a natural way gives rise to a one-good production model.

To this end consider an economy where consumers live for one period only. The number of consumers at date t, assumed to be equal to the available amount of labour at date t, is L_t. At date 1 , the first date, there is given an amount Y_1 of the good. Let $F: R_+^2 \to R_+$ be a function describing the production possibilities. The good may be consumed at date 1 or used as capital to produce, together with the available supply of labour, $F(K_1,L_1)$ units of the good which is then available at date 2. Here K_1 denotes the amount of capital used. The story is repeated at the dates 2,3,... and we get an evolution which may be visualized as follows

Date	1	2	3	t	t+1
Labour	L_1	L_2	L_3	L_t	L_{t+1}
Capital	K_1	K_2	K_3	K_t	K_{t+1}
Output	Y_1	Y_2	Y_3	Y_t	Y_{t+1}
Consumption	C_1	C_2	C_3	C_t	C_{t+1}

A <u>feasible program</u> is a sequence $(Y_t,K_t,L_t)_{t\varepsilon N}$ satisfying the following conditions for $t\varepsilon N$,

(i) Y_t, K_t and C_t are non-negative and $Y_t=K_t+C_t$

(ii) $Y_{t+1} \leq F(K_t,L_t)$

Assume that F is homogenous of degree 1 and define per capita variables $y_t=Y_t/L_t$, $x_t=K_t/L_t$, and $c_t=C_t/L_t$, for $t\varepsilon N$. The conditions (i) and (ii) are equivalent to the following conditions: For $t\varepsilon N$

(i') y_t, x_t and c_t are non-negative and $y_t=x_t+c_t$

(ii') $y_{t+1} \leq \frac{L_t}{L_{t+1}} F(x_t,1)=f_t(x_t)$

where f_t is defined by (ii'). f_t is independent of time if and only if L_{t+1}/L_t is constant over time.

Distinguishing goods according to the dates at which they are available it is seen that the above model involves infinitely many goods and could therefore be called an "infinite model". By contrast we shall occasionally refer to "finite models" or "finite economies", which will be used to denote models or economies with finitely many goods, like the Arrow-Debreu model(cf. Debreu [1959]).

Production efficiency may in the Swan-Phelps model and in the derived model be defined as for finite models and it is easy to see that there is a natural one-to-one correspondence between feasible programs in the original model and in the derived one. This correspondence maps efficient programs to efficient programs. In fact the transition to per capita variables may be interpreted simply as a rescaling of the amount of labour so as to give us precisely one unit of labour at each date. Since our primary interest is efficiency rather than the Swan-Phelps model we proceed with the model $(f_t)_{t\in N}$ and do not relate the following properties, assumed for f_t, to the properties of F.

ASSUMPTION 1.1. For $t\in N$, $f_t: R_+ \longrightarrow R_+$ is concave, non-decreasing and $f_t(0)=0$.

A <u>production model</u> is a sequence of functions $(f_t)_{t\in N}$ satisfying Assumption 1.1. A <u>feasible program</u> $(x_t, y_t, c_t)_{t\in N}$ for the production model $(f_t)_{t\in N}$ is a program satisfying assumptions (i') and (ii') above. A feasible program $(x_t', y_t', c_t')_{t\in N}$ <u>dominates</u> another feasible program if $c_t' \geq c_t$ for $t\in N$ with at least one strict inequality. A program is (production) <u>efficient</u> if it is feasible and there is no dominating program.

EXAMPLE 1.2. Let $L_t=1$, $L_{t+1}=3L_t$ and $F(K,L)=K^{(1/3)}L^{(2/3)}$. Then $(L_t/L_{t+1})F(x_t,1)=3x_t^{1/3}=f_t(x_t)$, $t\in N$,and $(f_t)_{t\in N}$ satisfies Assumption 1.1●

1.1.1 Price Supported Programs

In a finite economy we know that profits for individual producers are maximal if and only if total profits are maximal(cf.Debreu [1959]).In the present context the individual producer t, $t\epsilon N$, is the producer acting at dates t and t+1. Furthermore in a finite economy, for every efficient program there is a price system such that total profits are maximal for this program. Do these results carry over to the one-good production model?

First of all let us consider the existence of a supporting price system. For a given $t\epsilon N$, $q>0$ and $r>0$ define $\pi_t(q,r,y,x)=qy-rx$; the profit of the action (x,y). If (x,y) is possible for producer t, that is, $y\leq f_t(x)$, and maximizes profit relative to q and r then $y=f_t(x)$ and $D\pi_t(x)=qDf_t(x)-r=0$ provided f_t is differentiable. On the other hand, since f_t is concave and non-decreasing, if $y=f_t(x)$ then there are non-negative prices q and r such that (x,y) maximizes profit among the possible productions.

Indeed, if q and r may be chosen positive, then using the homogeneity of π_t in the prices, r may be chosen equal to 1 and then q serves as the relative price of output at date t+1 for producer t. The normalized price of input at date t+1 is, however, equal to 1 for producer t+1. To get a single price at date t let $(1,q_{t+1})$ support (x_t,y_{t+1}), put $q_1=1$ and define $p_t=\Pi_{i=1}^{t}q_i$. The price system $(p_t)_{t\epsilon N}$ so defined supports all the actions (x_t,y_{t+1}), $t\epsilon N$. Note that the case where $(x_t,y_t,c_t)_{t\epsilon N}$ is a feasible program with $x_t=0$ for some t is of little interest since $x_t=0$ will then hold except for finitely many $t\epsilon N$, in which case we are essentially back to the finite case.

Since f_t is assumed non-decreasing, rather than increasing, there may be productions satisfying $y=f_t(x)$ which can not be supported by positive prices.

Thus we are led to define a supporting price system for the feasible program $(\bar{x}_t,\bar{y}_t,\bar{c}_t)_{t\epsilon N}$ as a sequence $(p_t)_{t\epsilon N}$ such that, for $t\epsilon N$, $p_t>0$ and

$$p_{t+1}\bar{y}_{t+1}-p_t\bar{x}_t \geq p_{t+1}f_t(x_t)-p_tx_t$$

for $x_t\epsilon R_+$. The program $(\bar{x}_t,\bar{y}_t,\bar{c}_t)_{t\epsilon N}$ is then <u>price supported</u> and we also say that (p_t,p_{t+1}) is a <u>support</u> for $(\bar{x}_t,f_t(\bar{x}_t))$ for $t\epsilon N$.

If f_t is differentiable for $t\epsilon N$ then, $p_t=\Pi_{i=1}^{t}(1/Df_t(x_t))$ for $t\epsilon N$.

1.1.2 Steady States and Efficiency

Among the feasible programs, the steady states are the simplest as they are, by definiton, constant over time. What can be said about the efficiency of such programs? Let $(x,y,c)=(x_t,y_t,c_t)$ for $t\varepsilon N$ be a steady state and assume, for the moment, that f_t is time independent, that is, $f_t=f$ for $t\varepsilon N$, and differentiable with $Df(x')<1$ for some $x'\varepsilon R$. It is easy to see that a support to (x,y,c) will have the form $p_t=(1/Df(x))^t p_1$ for $t\varepsilon N$. Figure 1.1 illustrates the situation for some steady states. We note that for any steady state (x,y,c) where $Df(x)\leq 1$ total profits are not finite; hence the principle of decentralization over time can, in general, not be maintained in the formulation used for finite economies.

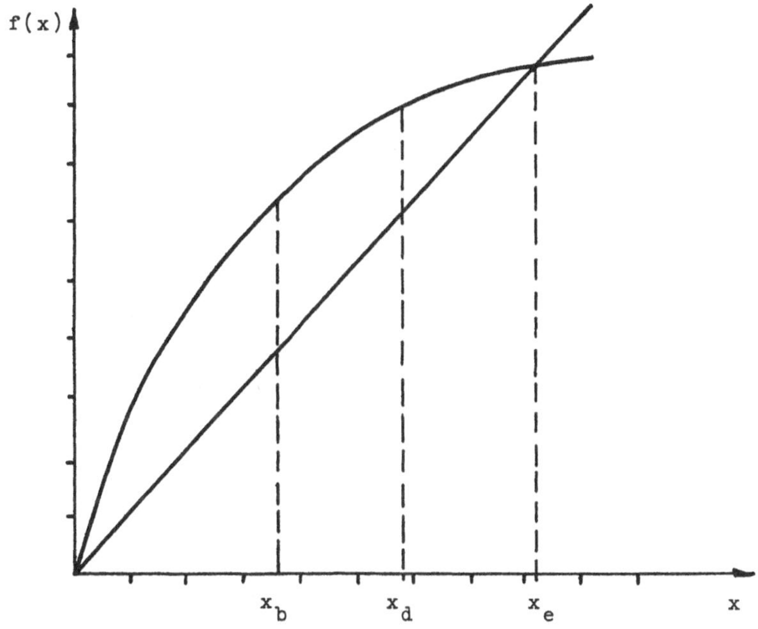

FIGURE 1.1

Clearly the steady state (x_d,y_d,c_d), $c_d=y_d-x_d$, is not efficient since the program defined by $c_1=c_d+(x_d-x_b)$, $x_1=x_b$, $y_1=y_d$ and $(x_t,y_t,c_t)=$ $=(x_b,y_b,c_b)$ for $t\geq 2$ gives $c_t\geq c_d$ for $t\varepsilon N$. Even worse than (x_d,y_d,c_d) is (x_e,y_e,c_e). Here all the resources are spent on maintaining the capital stock and consumption is $c_e=y_e-x_e=0$ in each period.

In the sequel we will show that a steady state (x,y,c) is efficient if and only if $Df(x)\geq 1$. The reason for the inefficiency of steady states where $Df(x)<1$ is that society "overaccumulates" capital, so that too much attention is given to having a large capital stock, in order to make possible large consumptions "at infinity", rather than in the future.

EXAMPLE 1.2. (continued). For $f_t=3x^{1/3}$, $t \in N$, we have for a steady state $y=3x^{1/3}=x+c$, that is,

$$x^{1/3}(3-x^{2/3})=c$$

Since $d3x^{1/3}/dx=(1/3)\cdot 3x^{-(2/3)}=x^{-(2/3)}$ we get maximal consumption for $x=1$. The different combinations of (x,c) possible for a steady state are shown in Figure 1.2. ●

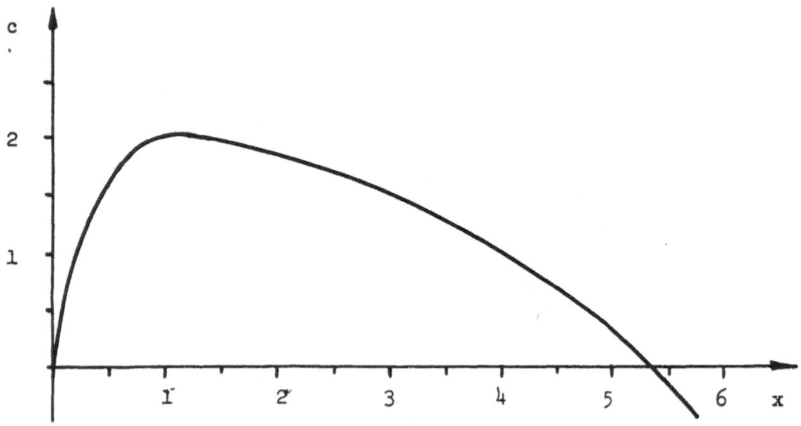

FIGURE 1.2

The failure of price supported programs to be, in general, efficient motivates the introduction of another efficiency concept. To give a rationale for this concept we ask whether it is possible for, say, the first hundred generations to improve the steady state (x_d,y_d,c_d) by, decreasing the capital stock in the beginning and correspondingly increase consumption and in the end restore capital to x_d so as to leave the generations to follow no worse off than before.

Our problem may, somewhat generalized, be rephrased as follows: Is there a program dominating (x_d,y_d,c_d) and differing from it for finitely many dates only? A feasible program $(x_t,y_t,c_t)_{t \in N}$ such that there is no dominating program differing from it for finitely many dates only is called a weakly efficient program. For a weakly efficient program we always have $y_{t+1}=f_t(x_t)$, $t\epsilon N$.

We propose to answer the question raised above in the negative by showing that any price supported program is weakly efficient. For any price system p_t and any feasible program we have for $t\epsilon N$,

$$p_tc_t+\ldots+p_1c_1=p_t(y_t-x_t)+\ldots+p_1(y_1-x_1)=$$

$$=\pi_{t-1}(p_t,p_{t-1},y_t,x_{t-1})+\ldots+\pi_1(p_2,p_1,y_2,x_1)+(p_1y_1-p_tx_t)$$

Assume that $(x_t,y_t,c_t)_{t\epsilon N}$ is supported by $(p_t)_{t\epsilon N}$. Let $(x_t',y_t',c_t')_{t\epsilon N}$ be a dominating program differing from the first one only for $t<T$. Then, using the equality above,

$$\Sigma_{t=1}^{T}p_tc_t'=\Sigma_{t=1}^{T-1}\pi_{t+1}(p_{t+1},p_t,y_{t+1}',x_t')+(p_1y_1'-p_Tx_T') >$$

$$> \Sigma_{t=1}^{T}p_tc_t=\Sigma_{t=1}^{T-1}\pi_{t+1}(p_{t+1},p_t,y_{t+1},x_t)+(p_1y_1-p_Tx_T)$$

The maximization of profits in each period implies that the sum of profits given by the left hand side in the inequality above is no greater than the corresponding sum on the right hand side. Hence, since $y_1'=y_1$ by feasibility, $p_Tx_T'<p_Tx_T$ which contradicts $x_T'=x_T$.

This shows that price supported programs are weakly efficient. They may, however, as shown by the examination of steady states, fail to be efficient. It follows that a program dominating a price supported program must differ from it for infinitely many dates.

1.1.3 Improving Sequences

A natural approach when trying to construct a dominating program is to try to increase consumption at some date; this can be achieved only by decreasing capital at that date and so, in turn, we get less consumption at the next date. Since we can not decrease consumption at that date - we want a dominating program - capital has to be decreased at least as much as the reduction in production and so on. When the initial program is only weakly efficient the process may not come to an end. If it does, it is only because at some date, the required reduction in capital would force capital to take on a negative value.

There is an alternative convenient way, suggested by the former reasoning, to express the fact that a given weakly efficient program is not efficient, which we now proceed to describe.

Let $(x'_t, y'_t, c'_t)_{t\in N}$ be a program dominating the weakly efficient program $(x_t, y_t, c_t)_{t\in N}$ so that we have

1	2	3		t	t+1
x_1	x_2	x_3		x_t	x_{t+1}
y_1	y_2	y_3		y_t	y_{t+1}
c_1	c_2	c_3		c_t	c_{t+1}

and

1	2	3		t	t+1
x'_1	x'_2	x'_3		x'_t	x'_{t+1}
y'_1	y'_2	y'_3		y'_t	y'_{t+1}
c'_1	c'_2	c'_3		c'_t	c'_{t+1}

and $c'_t \geq c_t$ for $t\in N$ with strict inequality for some t. Define $(\xi_t)_{t\in N}$ by $\xi_t = x_t - x'_t$ for $t\in N$. Then $\xi_1 = c'_1 - c_1$ and

$$\xi_{t+1} = (y_{t+1} - c_{t+1}) - (y'_{t+1} - c'_{t+1}) = f_t(x_t) - f_t(x_t - \xi_t) +$$
$$+ (f_t(x_t - \xi_t) - y'_{t+1}) + (c'_{t+1} - c_{t+1}) \geq f_t(x_t) - f_t(x_t - \xi_t)$$

for $t\in N$. Since $y'_{t+1} \leq f_t(x'_t) = f_t(x_t - \xi_t)$ and $c'_t - c_t \geq 0$ for $t\in N$ we get $\xi_t \geq 0$ and for t such that $c'_t - c_t > 0$ we have $\xi_t > 0$ since the original program was assumed weakly efficient and f_t is non-decreasing.

Guided by the relations satisfied by the sequence $(\xi_t)_{t\in N}$ we now define an improving sequence, $(\xi_t)_{t\in N}$, for a weakly efficient program $(x_t, y_t, c_t)_{t\in N}$ as a sequence satisfying, for $t\in N$,

(i) $\xi_t \geq 0$ and $\xi_t > 0$ for some $t\in N$

(ii) $x_t - \xi_t \geq 0$

(iii) $\xi_{t+1} \geq f_t(x_t) - f_t(x_t - \xi_t)$

If (iii) is always satisfied with equality, corresponding to $c_t = c'_t$ except for precisely one t, then we call the improving sequence simple.

Thus a dominating program induces an improving sequence. The converse is also true; let $(x_t,y_t,c_t)_{t\in N}$ be a weakly efficient program with $(\xi_t)_{t\in N}$ an improving sequence. Then

1	2	t	t+1
$x_1-\xi_1$	$x_2-\xi_2$	$x_t-\xi_t$	$x_{t+1}-\xi_{t+1}$
y_1	$f_1(x_1-\xi_1)$	$f_{t-1}(x_{t-1}-\xi_{t-1})$	$f_t(x_t-\xi_t)$
$c_1+\xi_1$	$f_1(x_1-\xi_1)-(x_2-\xi_2)$	$f_{t-1}(x_{t-1}-\xi_{t-1})-$ $-(x_t-\xi_t)$	$f_t(x_t-\xi_t)-$ $-(x_{t+1}-\xi_{t+1})$

is a dominating program since $c_1+\xi_1\geq c_1$ and consumption at date t+1 equals

$$f_t(x_t-\xi_t)-(x_{t+1}-\xi_{t+1})=$$

$$=(\xi_{t+1}-(f_t(x_t)-f_t(x_t-\xi_t))+(f_t(x_t)-x_{t+1})\geq c_{t+1}$$

for $t\in N$ and we get strict inequality for $t\geq\min\{t\,|\,\xi_t>0\}$.

We do not know very much about the improving sequence itself but if the considered program is supported by, say, $(p_t)_{t\in N}$, then $(p_t\xi_t)_{t\in N}$ is always non-decreasing since

$$p_{t+1}\xi_{t+1}-p_t\xi_t\geq p_{t+1}(f_t(x_t)-f_t(x_t-\xi_t))-p_t\xi_t=$$

$$=(p_{t+1}f_t(x_t)-p_tx_t)-(p_{t+1}f_t(x_t-\xi_t)-p_t(x_t-\xi_t))\geq 0$$

Here we used that the production $(x_t,f_t(x_t))$ is profit maximizing. We note that if f_t is linear ($f_t=a_tx$, $a_t>0$ for $t\in N$) and if the improving sequence is simple then, as can be seen from the relation above, $p_t\xi_t$ is non-negative and constant for all $t\geq\min\{t\,|\,\xi_t>0\}$.

EXAMPLE 1.2.(continued). Consider the steady state defined by $x=2$, $y=3\cdot 2^{1/3}$ and $c=3\cdot 2^{1/3}-2$. Let $(x_t,y_t,c_t)_{t\in N}$ be defined by $x_1=1$, $c_1=3\cdot 2^{1/3}-1$ and for $t>1$, $y_t=3$, $x_t=1$ and $c_t=2$.

The corresponding improving sequence is $\xi_1=c_1-c=1$ and $\xi_t=x-x_t=1$. A support for the original program is given by $p_t=(2^{(2/3)})^t$ showing that $p_t\xi_t$ is non-decreasing. ●

1.1.4 Efficiency Criteria

Since, for an improving sequence, $x_t-\xi_t\geq 0$, or equivalently $p_t x_t \geq p_t \xi_t$, and $p_t\xi_t$ is non-decreasing there can be no improving sequence if liminf $p_t x_t=0$. This sufficient condition for efficiency, liminf $p_t x_t=0$, is due to Malinvaud [1953] who proved it in a more general context. We leave it to the reader to check that in the linear case this condition is also necessary; if liminf $p_t x_t>0$ then there exists an improving sequence. This was first pointed out by McFadden [1967].

When f_t is linear and independent of time ($f_t(x)=ax$ for $t\varepsilon N$) and $a=1$ we have a pure storage process. Imagine a cake, parts of which can be consumed at different dates, and where the leftovers can be stored until the next date. A program will be efficient if and only if no part of the cake is stored forever.

Malinvaud's criterion was the first in a series of criteria for efficiency or inefficiency in infinite horizon models. The inequality $p_t x_t \geq p_t \xi_t$ suggests that it is the "growth rate" in $p_t x_t$ as compared to $p_t\xi_t$ that is of interest. The criteria obtained subsequently by Cass [1972a] and Benveniste and Gale [1975] may be interpreted to say that a program is efficient if and only if $p_t x_t$ does not grow "too fast". These criteria, as well as that of Mitra [1979b], were derived by imposing conditions which restricted the rate of growth of any improving sequence.

Cass showed that the series $\Sigma_{t\varepsilon N}(1/p_t)$ is convergent if and only if the program is inefficient. Although Cass' result was restricted to programs satisfying a uniform boundeness condition, $\bar{x}\leq x_t \leq x''$ for some $0<\bar{x}<x''$, and was proved for the case where f_t is independent of time and twice continuously differentiable, the result is quite useful as the reader can see by applying it to the steady states discussed earlier. Furthermore, the method of proof influenced the subsequent contributions to a large extent (cf. the calculations in Example 1.5, Section 1.2.7).

Benveniste and Gale [1975] pointed out that Cass' result could easily be generalized to allow for time dependent production possibilities. Once this is allowed for there is no compelling reason to use the same units of measurement at different dates. If we measure in kilos at date 1 we may measure in pounds at date 2 and so on. In fact, following the tradition of general equilibrium theory, the goods at different dates may bear no similarity whatsoever. Cass' criterion is not invariant for changes in the units of measurement since these entail corresponding changes in the prices. On the other hand, Cass' assumptions does not allow for arbitrary changes in the units of measurement.

Benveniste and Gale [1975] showed that a program is inefficient if and only if $\Sigma_{t=1}^{\infty}(1/p_t x_t)$ is convergent; this latter sum being invariant for changes in the units of measurement. Note that when $x_t=1$ for $t\varepsilon N$ this criterion coincides with that of Cass.

Benveniste and Gale used in their proof the fact that units may be chosen so that $x_t=1$ for $t\varepsilon N$. Varying the units of measurement can be a useful approach in several ways. It may be used to simplify proofs, to get a better understanding of the problem at hand, and it may draw our attention to assumptions which are not invariant under such changes.

The alternative approach to criteria for efficiency is to focus on the growth of $p_t\xi_t$. Figure 1.3 exhibits the situation when we try to define a (simple) improving sequence

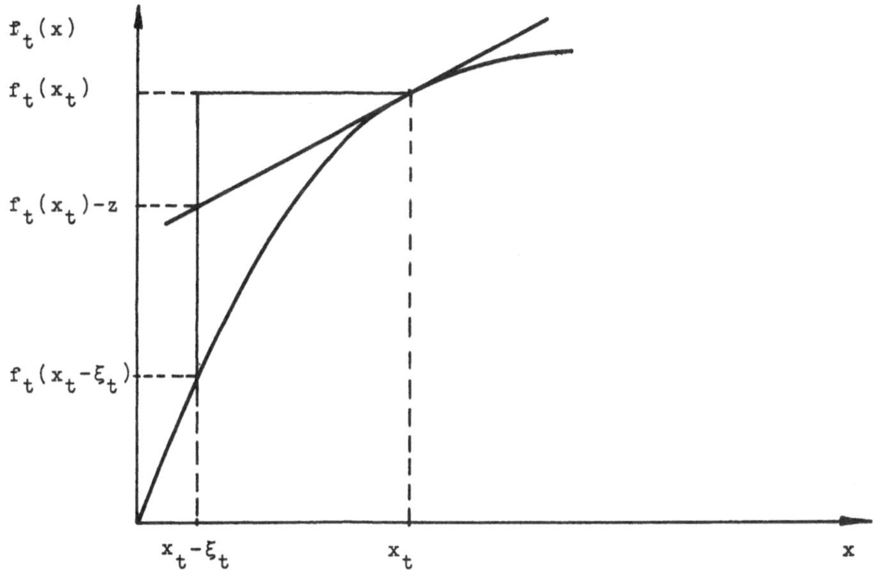

FIGURE 1.3

If the graph of f_t had been the line tangent at $(x_t,f_t(x_t))$ we would have had $p_{t+1}\xi_{t+1}=p_t\xi_t$ and $\xi_{t+1}=f_t(x_t)-z$. But this is not the case and we get instead

$$p_{t+1}\xi_{t+1}=p_{t+1}(f_t(x_t)-z)+p_{t+1}(z-f_t(x_t-\xi_t))=p_t\xi_t+p_{t+1}(z-f_t(x_t-\xi_t))$$

which shows that $p_t\xi_t$ grows faster the greater the "curvature" of f_t is.

We know from Malinvaud's criterion that if $\liminf p_t x_t=0$, then the program is efficient. If the "curvatures" at $(x_t, f_t(x_t))$, for $t\varepsilon N$, are small and $\liminf p_t x_t>0$, one would suggest that there exists an improving sequence, in particular if it were possible to prove, under the conditions given, that there will always be a sequence $(\xi_t)_{t\varepsilon N}$ such that $p_t\xi_t$ is bounded. However this is not so; we shall return to these problems in the chapters to follow.

EXAMPLE 1.2. (continued). Consider a program $(x_t,y_t,c_t)_{t\varepsilon N}$ that oscillates around the "golden rule" program; $(x,y,c)=(3,1,1)$. Let

$$x_t=\begin{cases}1/8 \text{ if } t \text{ is odd}\\ 9/8 \text{ if } t \text{ is even}\end{cases}$$

Then, as some calculations show,

$$y_t=\begin{cases}12/8 & \text{if } t \text{ is odd}\\ (12/8)\cdot 9^{1/3} & \text{if } t \text{ is even}\end{cases}$$

$$c_t=\begin{cases}3/8 & \text{if } t \text{ is odd}\\ (1/8)\cdot(12\cdot 9^{1/3}-1) & \text{if } t \text{ is even}\end{cases}$$

so we have the following situation

	1	2	2t	2t+1
x_t	1/8	9/8	1/8	9/8
y_t	$(12/8)\cdot 9^{1/3}$	12/8	$(12/8)\cdot 9^{1/3}$	12/8
c_t	$(1/8)(12\cdot 9^{1/3}-1)$	3/8	$(1/8)(12\cdot 9^{1/3}-1)$	3/8
p_t	1	1/4	$\frac{(9^{2/3})^{t-1}}{4^{2t-1}}$	$\frac{(9^{2/3})^{t}}{4^{2t}}$

The last line shows a supporting price system where $p_1=1$, and

$$p_{2t}=\frac{(9^{2/3})^{t-1}}{4^{2t-1}}, \quad p_{2t+1}=\frac{(9^{2/3})^{t}}{4^{2t}} \qquad \text{for } t\geq 1$$

The assumptions needed to apply the criterion of Cass or Benveniste and Gale are satisfied. We have

$$\Sigma_{t=1}^{\infty}(1/p_t)=1+\Sigma_{t=1}^{\infty}(1/p_{2t})+\Sigma_{t=1}^{\infty}(1/p_{2t+1}) \geq \Sigma_{t=1}^{\infty}[\,4^{2t}/(9^{2/3})^t]$$

which shows that the program is efficient by Cass' criterion. Benveniste and Gale's criterion gives the same result and in fact liminf $p_t x_t=0$, so the program is efficient already by Malinvaud's criterion. ●

1.2. THE SAMUELSON CONSUMPTION MODEL

In the production model of Section 1.1 the only interdependence between dates was that the output at a certain date originated from input applied at the previous date. We will show now that we might introduce consumers, one for each date, so that a price supported program can be regarded as an equilibrium for consumers and producers.

Given the program $(x_t,y_t,c_t)_{t\varepsilon N}$ let consumer t, for t>2, be defined by initial resources $(\omega_{t-1},\omega_t)=(x_{t-1},-x_t)$ and consumption set $X_t=\{c_t|c_t>0\}$. Putting the initial resources for consumer 1 equal to $\omega_1=c_1$ we get $p_1c_1=p_1\omega_1$ and, for $t\geq 2$,

$$p_tc_t=p_t(y_t-x_t)=(p_tf_{t-1}(x_{t-1})-p_{t-1}x_{t-1})-p_tx_t+p_{t-1}x_{t-1}=$$

$$=\pi_{t-1}(p_t,p_{t-1},y_t,x_{t-1})+p_t\omega_t+p_{t-1}\omega_{t-1}$$

showing that the consumptions c_t are possible with respect to the consumers' budgets when consumer t, $t\geq 2$, is the sole owner of producer t-1. The consumption c_t will, of course, be an equilibrium consumption provided that consumers have strictly monotone preferences.

We shall now consider a related model, the celebrated overlapping generations model, introduced by Samuelson [1958]. This model is more complicated than the previous one, since every consumer (except the first one) consumes at two consecutive dates but at the same time it is simpler since there is no production.

1.2.1. The Overlapping Generations Model

For future reference we introduce a model which is a bit more general than needed in this chapter and then introduce some simplifying assumptions.

We consider a model where for $t\in N\cup\{0\}=M$ there is a generation born at t which consumes at date t+1 and, if $t\in N$, at date t. $M(t)=\{(t,i)|t=\tau\}$ is the generation born at τ for $\tau\in M$

We shall use the convention that any member of generation t lives at dates t,t+1, for $t\geq 1$; the initial generation lives only at date 1.

Clearly, what matters in our model are not the physiological aspects of life; rather the main point is that each consumer wants to trade in, at most, two consecutive time periods, thereby assuring that for every single time period there is only a finite number of consumers which are actually trading.

We make the following:

ASSUMPTION 1.3.
(i) For $i\in M(t)$, $t\in N$, $X^{t,i}$, the consumption set of consumer (t,i), is R^2_+. The initial resources of (t,i), $\omega^{t,i}$, belong to $\text{int}R^2_+$ and the preference correspondence $x^{t,i}\rightarrow P^{t,i}(x^{t,i})\subset R^2_+$, $x^{t,i}\in X^{t,i}$, is irreflexive, satisfies $P^{t,i}(x^{t,i})+R^2_+\subset P^{t,i}(x^{t,i})$, has open graph and is convex valued.

(ii) For $i\in M(0)$, $X^{0,i}=R_+$, $\omega^{0,i}\in R_+$ and $P^0(x^{0,i})=\{x|x>x^{0,i}\}$ for $x^{0,i}\in R_+$

In this chapter we will assume furthermore that for $t\in N$, there is only one consumer, denoted t, and that his preference correspondence is generated by a utility function U^t: $R^2_+\rightarrow R$. A typical consumption for $t\in N$ will be denoted $x^t=(x^t_F,x^t_L)$ (First and Last date) and his initial resources are (ω^t_F,ω^t_L). For $t\in N$ the consumer is then described by

(R_+^2, U^t, ω^t) and for $t=0$, by (R_+, U^0, ω^0) where $U^0: R_+ \to R$ and $\omega^0 = \omega_L^0$. A typical consumption for consumer 0 is $x^0 = x_L^0$.

A <u>program</u> in the economy is a sequence $(x^t)_{t\epsilon M}$ of consumptions. It can be visualized as follows:

1	2	3		t	t+1
x_F^1	x_F^2	x_F^3		x_F^t	x_F^{t+1}
x_L^0	x_L^1	x_L^2		x_L^{t-1}	x_L^t

The corresponding schedule of initial resources determines the supply of the good at each date. The program, $(x^t)_{t\epsilon m}$, is <u>feasible</u> if for $t\epsilon N$

$$x_L^{t-1} + x_F^t = \omega_L^{t-1} + \omega_F^t, \quad x_L^0 \epsilon R_+ \text{ and } x^t \epsilon R_+^2.$$

1.2.2. Equilibrium Programs

A feasible program $(x^t)_{t\epsilon M}$ <u>dominates</u> the feasible program $(\bar{x}^t)_{t\epsilon M}$ if $U^t(x^t) \geq U^t(\bar{x}^t)$ for $t\epsilon M$ with at least one strict inequality. As in a finite economy a feasible program is <u>Pareto optimal</u> if there is no program dominating it. A <u>price system</u> is a sequence of positive prices $(p_t)_{t\epsilon N}$. Let $p^t = (p_F^t, p_L^t) = (p_t, p_{t+1})$ for $t\epsilon N$ and $p_L^0 = p_1$; these are the prices of interest to consumer $t\epsilon M$.

A feasible program $(\bar{x}^t)_{t\epsilon M}$ is <u>price supported</u> if for some price system $(p_t)_{t\epsilon N}$, for $t\epsilon M$,

(i) $U^t(x^t) > U^t(\bar{x}^t)$ implies $p^t x^t > p^t \bar{x}^t$

The program is a (private ownership) <u>equilibrium</u> if it is price supported and

(ii) $p_1 \bar{x}_L^0 \geq p_1 \omega_L^0$ and, for $t\epsilon N$, $p^t \bar{x}^t = p_F^t \bar{x}_F^t + p_L^t \bar{x}_L^t \geq p^t \omega^t$

In relation to condition (ii) we note that $p_1 \omega_L^0 + \Sigma_{t\epsilon N} p^t \omega^t$ and $p_1 x_L^0 + \Sigma_{t\epsilon N} p^t x^t$ will, in general, not be finite. Therefore Walras' law, valid in a finite economy and stating that for any price system, the value of demand will equal the value of initial resources, does not have an immediate counterpart in infinite economies. Example 1.4 below exhibits an equilibrium where (ii) is satisfied with strict inequality for each consumer.

It is comforting to know that the failure of Walras' law to hold does not preclude the existence of equilibria. Existence proofs have been given by Balasko, Cass and Shell [1980], Borglin and Keiding [1981] and Wilson [1981] under assumptions similar to those used for a finite economy. The basic idea has been to show the existence of equilibrium for finite submodels and obtain an equilibrium for the infinite model as a limit of such equilibria.

1.2.3. Weak Optimality and Optimality

Will a price supported program $(\bar{x}^t)_{t\epsilon M}$ necessarily be Pareto optimal? Let $(p^t)_{t\epsilon N}$ be a support to the program and assume that there is a dominating program $(x^t)_{t\epsilon M}$. Then

$$p_1 x_L^0 \geq p_1 \bar{x}_L^0 \text{ and } p^t x^t \geq p^t \bar{x}^t \text{ for } t\epsilon N$$

with strict equality at least for one t, say, $t=\tau$. For any $T>\tau$ we have, by summation,

$$0 < (p_1 x_L^0 + \Sigma_{t\epsilon N} p^t x^t) - (p_1 \bar{x}_L^0 + \Sigma_{t\epsilon N} p^t \bar{x}^t) =$$

$$= \Sigma_{t=T+1}^{\infty} (p^t x^t - p^t \bar{x}^t) + p_T (x_L^T - \bar{x}_L^T)$$

where the last equality is derived by repeated use of the feasibility conditions.

Assume that there is T_o such that $x^t = \bar{x}^t$ for $t > T_o$. Then

$$\Sigma_{t=T+1}^{\infty} (p^t x^t - p^t \bar{x}^t) + p_T (x_L^T - \bar{x}_L^T)$$

is 0 for T sufficiently large, contradicting the inequality above. Thus a price supported program is <u>weakly optimal</u>; there can be no dominating program differing from the given one for finetely many dates only. The reasoning above also indicates why a price supported program may fail to be Pareto optimal. There is no contradiction in assuming that there is a program giving each of an infinite number of consumers a consumption more valuable than their original one.

Any program having a support is thus seen to be weakly optimal. The converse, asserting that a weakly optimal program has a support is also true provided that there are positive prices, $\bar{p}^t=(\bar{p}_F^t,\bar{p}_L^t)$ for $t\in N$, supporting the individual consumptions of the program. In this case we may define $(p_t)_{t\in N}$ by $p_1=1$, $p_2=\bar{p}_L^1/\bar{p}_F^1$ and, in general, $p_{t+1}=(\bar{p}_L^t/\bar{p}_F^t)p_t$.

1.2.4. Improving Sequences

The preceding sections show that the situation is similar to the one for the production model where the price supported programs were shown to coincide with the weakly efficient programs. For the production model there was a convenient way to describe a program dominating a (weakly efficient) program. We will now show that something similar can be done in the consumption model. Let $(\bar{x}^t)_{t\in M}$ be a weakly optimal program with support $(p_t)_{t\in N}$.

1	2	3		t	t+1
$\bar{x}_F^1$	$\bar{x}_F^2$	$\bar{x}_F^3$		$\bar{x}_F^t$	$\bar{x}_F^{t+1}$
$\bar{x}_L^0$	$\bar{x}_L^1$	$\bar{x}_L^2$		$\bar{x}_L^{t-1}$	$\bar{x}_L^t$

If a program is to be dominating then obviously some consumer must receive more of some good. Assume that consumer 0 receives more. The only consumer who can supply good 1 to consumer 0 is consumer 1. To keep consumer 1 at least as well of as before he has to be compensated in good 2 and this can be achieved only through compensation by consumer 2. The process may be continued indefinitely unless , for some $t\in N$,it is impossible for consumer t to make the required reduction in his first date consumption; that is,

$$\bar{x}_F^t-\xi_t \geq 0$$

and

$$U^{t-1}(\bar{x}_F^{t-1}-\xi_{t-1},\bar{x}_L^{t-1}+\xi_t) \geq U^{t-1}(\bar{x}_F^{t-1},\bar{x}_L^{t-1})$$

lacks a solution in ξ_t, where $\xi_t>0$, $t\epsilon N$, denotes the amount that consumer t delivers to t-1. Assuming that the process may always be continued we get

1	2	3		t	t+1
$\bar{x}_F^1-\xi_1$	$\bar{x}_F^2-\xi_2$	$\bar{x}_F^3-\xi_3$		$\bar{x}_F^t-\xi_t$	$\bar{x}_F^{t+1}-\xi_{t+1}$
$\bar{x}_L^0+\xi_1$	$\bar{x}_L^1+\xi_2$	$\bar{x}_L^2+\xi_3$		$\bar{x}_L^t+\xi_t$	$\bar{x}_L^t+\xi_{t+1}$

It is now clear how an improving sequence should be defined; $(\xi_t)_{t\epsilon N}$ is an <u>improving sequence</u> for the weakly optimal program $(\bar{x}^t)_{t\epsilon M}$ if, for $t\epsilon N$,

(i) $\xi_t \geq 0$ and $\xi_t> 0$ for some t

(ii) $\bar{x}_F^t-\xi_t \geq 0$ for $t\epsilon N$

(iii) $U^t(\bar{x}_F^t-\xi_t,\bar{x}_L^t+\xi_{t+1}) \geq U^t(\bar{x}^t)$ for $t\epsilon N$ and $U^0(x_F^0+\xi_1)\geq U^0(x_F^0)$

By strict monotonicity of preferences we will get at least one strict inequality in (iii). If equality holds in (iii) except for one date we call the improving sequence <u>simple</u>. The improving sequence $(\xi_t)_{t\epsilon N}$ induces the dominating program shown above. On the other hand, if $(x^t)_{t\epsilon M}$ dominates the weakly efficient program $(\bar{x}^t)_{t\epsilon M}$ then $\xi_t=\bar{x}_F^t-x_F^t$, $t\epsilon N$, is an improving sequence.

1.2.5. Optimality Criteria

Let $(p_t)_{t\epsilon N}$ be a supporting price system and $(\xi_t)_{t\epsilon N}$ an improving sequence to the weakly efficient program $(\bar{x}^t)_{t\epsilon M}$. Then

$$p_t(\bar{x}_F^t-\xi_t)+p_{t+1}(\bar{x}_L^t+\xi_{t+1}) \geq p_t\bar{x}_F^t+p_{t+1}\bar{x}_L^t \text{ for } t\epsilon N.$$

Hence $p_1\xi_1\geq 0$ and $p_{t+1}\xi_{t+1}\geq p_t\xi_t$ for $t\epsilon N$ which shows that $p_t\xi_t$ is non-decreasing.

Malinvaud's criterion, cf. Section 1.1.5, carries over to the consumption model; if liminf $p_t \bar{x}_F^t = 0$ then there can be no improving sequence. Balasko and Shell [1980] have shown that under rather restrictive assumptions, the Cass criterion also carries over: a program is efficient if and only if $\Sigma_{t \in N}(1/p_t)$ is convergent.

Weakly optimal programs may also be characterized by the marginal conditions satisfied by such programs. For an interior weakly optimal program $(x^t)_{t \in M}$ the marginal conditions, well known from finite economies, will be satisfied. Thus there are prices $(p_t)_{t \in N}$ and numbers $(\alpha_t)_{t \in N}$ such that $(DU_F^t(x^t), DU_L^t(x^t) = \alpha_t(p_t, p_{t+1})$ for $t \in N$.

As in a finite economy these conditions indicate the impossibility of obtaining something preferred within the budget determined by the prices and the given consumption. This in turn implies the absence of any mutually advantageous exchanges between any finite number of participants in the market.

However, we know that this does not automatically entail optimality in the infinite horizon model. The crucial difference between this model and the finite one is that, in order to check for optimality in the finite model, we never need to consider variations of the budget. Walras' law holds, and it can be interpreted to state that there is a strict conflict between budgets; one consumer can get a more expensive consumption only if someone else gets a cheaper one. In the infinite model this conflict may not be present for a price supported program and it may be possible, as has been argued above, to find mutually advantageous exchanges allowing each of an infinite number of consumers to enjoy a consumption giving higher utility and more valuable than his original one.

1.2.6. An Interpretation

The interpretation of prices carry over from the finite model. We may imagine an auctioneer calling out prices at date 0 and consumers responding with their demand. The price p_t is the number of units of account that is paid at date 0 to (by) the consumer for the delivery of (to have delivered) one unit of the good at date t. Since, by the conditions for equilibrium, incomes do not necessarily equal each of the values of initial resources, the auctioneer may also call incomes.

To respond, consumer t needs to know only his income and p_t. It is noteworthy that, in spite of the sequential structure of the model, all prices have to be determined simultaneously. Trying to find prices equilibrating the first T markets, we are forced to consider the first T consumers who consume the first T+1 goods, that is, we need T+1 prices. To make sure that p_{T+1} is chosen correctly we have to include consumer T+1, who consumes goods T+1, T+2 and so on ad infinitum.

There are, of course, other interpretations of the finite model which also carry over to the consumption model considered here.

1.2.7. Two Examples

We conclude this section with two examples. The first one shows that there are equilibria where Walras' law does not hold and that, if we insist on equilibria where incomes equal the value of initial resources, then there may be a considerable loss in consumer satisfaction. The second example introduces a consumption model which is of central importance for the considerations of efficiency/optimality in the general situation. At this stage, however, we consider it only as an illustration of the concepts introduced above and as an example of the method introduced by Cass (cf. Section 1.1.5).

EXAMPLE 1.4. Put $\omega_L^0=1$ and, for $t\in N$, let $\omega^t=(2^t,2^t)$, $U^t(x^t)=x_F^t+x_L^t$. Then $p_t=1$, $\bar{x}_L^0=\omega_L^0$ and $\bar{x}^t=\omega^t$, for $t\in N$, defines an equilibrium where $p^t\bar{x}^t=p^t\omega^t$. On the other hand, $x_L^0=2$, and for $t\in N$,

$$(x_F^t,x_L^t)=(2^t-2^{t-1},2^t+2^t)=(2^{t-1},2^{t+1})$$

is another equilibrium, relative to the same price system. Since

$$U^t(x^t)=p^tx^t=2^{t+1}\,\frac{5}{4} > 2^{t+1} = U^t(\bar{x}^t),$$

for $t\in N$ and $x_L^0=2>\omega_L^0=\bar{x}_L^0=1$ the original equilibrium is not Pareto optimal. ●

EXAMPLE 1.5. We define a consumption model by

$$U^0(x_L^0)=x_L^0$$

$$U^t(x^t)=(1-a_t)x_F^t+(1-a_t)x_L^t+a_t x_F^t x_L^t$$

and $\omega_L^0=1$, $\omega^t=(2,0)$, for $t\in N$, where $(a_t)_{t\in N}$ is a sequence of numbers with $a_1=1$ and $0<a_t\leq 1$ for $t\geq 2$. A typical consumer $t\in N$ is depicted in Figure 1.4, where we have also drawn a budget line corresponding to prices $p^t=(p_t,p_{t+1})=(1,1)$. The indifference curves are translated hyperbola with a curvature depending on the parameter a_t.

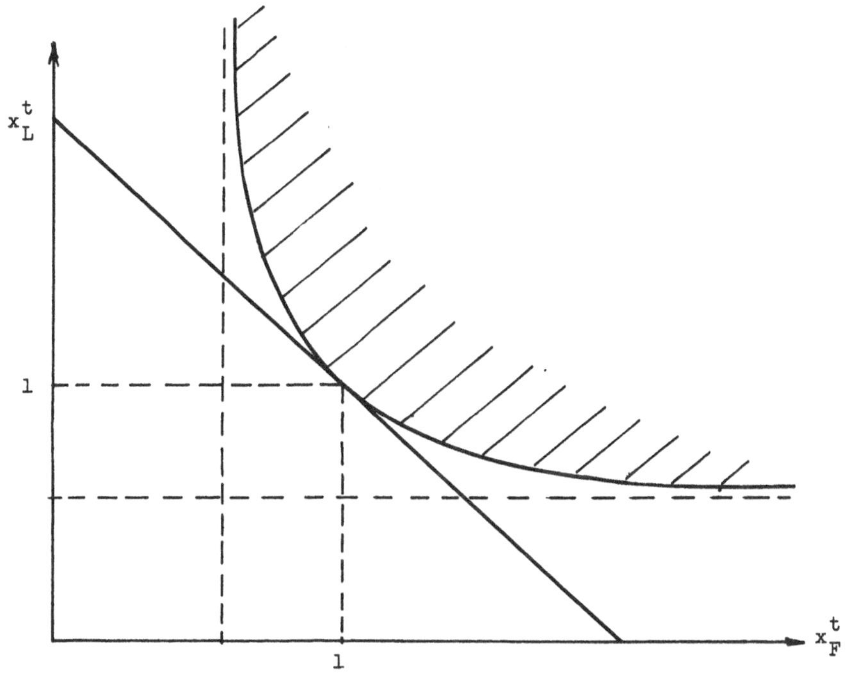

FIGURE 1.4

It is easily checked that for each $t\in N$, the consumption $x^t=(1,1)$ is supported by $p_t=(p_t,p_{t+1})$. Actually, the program $(x^t)_{t\in M}$ defined by $x_L^0=1, x^t=(1,1)$ for $t\in N$, is an equilibrium. Is it Pareto optimal?

First of all, we note that $\liminf p_t x_F^t=1>0$ so we can not conclude anything from the Malinvaud criterion. Since $\Sigma_{t=1}^{\infty}(1/p_t)=\infty$ the Balasko-Shell criterion would yield that the program is Pareto optimal, but the assumptions behind this criterion are not satisfied (for $(a_t)_{t\in N}$ an arbitrary sequence, the Gaussian curvature of the sets $\{x|U^t(x)\geq U^t(1,1)\}$ need neither be upper nor lower bounded). This could be remedied by a change of units of measurement. However, a direct approach is easier and more instructive.

Suppose that consumer 1 was to give up $\xi_1>0$ to consumer 0 (whose utility undoubtedly increases by this transaction). To what extent must he be compensated in good 2 in order that

$$U^1(x_F^1,x_L^1)=U^1(x_F^1-\xi_1,x_L^1+\xi_2)$$

Clearly, the solution must satisfy

$$(1-\xi_1)(1+\xi_2)=1$$

or

$$\xi_2-\xi_1=\xi_1\xi_2$$

or

$$\frac{1}{\xi_1}-\frac{1}{\xi_2}=1$$

Similarly we can see that if consumer 1 is compensated by consumer 2 to the amount of ξ_2, then consumer 2 must be given at least an amount ξ_3 satisfying

$$(1-a_2)(1-\xi_2)+(1-a_2)(1+\xi_3)+a_2(1-\xi_2)(1+\xi_3)=2-a_2$$

or

$$\frac{1}{\xi_2}-\frac{1}{\xi_3}=a_2$$

Proceeding in this way, we get for each t that

$$\frac{1}{\xi_t}-\frac{1}{\xi_{t+1}}=a_t$$

and consequently

$$\Sigma_{t=1}^{\tau}\left(\frac{1}{\xi_t}-\frac{1}{\xi_{t+1}}\right)=\left(\frac{1}{\xi_1}-\frac{1}{\xi_{t+1}}\right)=\Sigma_{t=1}^{\tau}a_t$$

Now if the sequence $(\xi_t)_{t\epsilon N}$ exists, then

$$\frac{1}{\xi_1} - \frac{1}{\xi_{t+1}} \le \frac{1}{\xi_1}$$

so the series $\Sigma_{t=1}^{\infty} a_t$ is convergent. Thus it is seen that if $\Sigma_{t=1}^{\infty} a_t$ is divergent then the program considered is Pareto optimal.

The converse is also true; if $\Sigma_{t=1}^{\infty} a_t$ is convergent then the program is not Pareto optimal. We shall not give a demonstration of this here; it will follow readily from the considerations in Chapter 3. ●

CHAPTER 2: REDUCED MODELS

In the first chapter we have given a brief treatment of the type of models and the kind of problems which we shall be studying throughout this book. In the present chapter we start the systematic treatment of these problems.

Our first step will be to transform the seemingly different production and consumption models to models belonging to a single class and containing all the information necessary for the efficiency/optimality considerations. Models of this type are called reduced models and are introduced in Section 1 below. In the following section we show how the reduction of the original production or consumption model can be carried out.

Reduced models are useful since they allow us us to concentrate upon the essential features of the problem. Their importance carries further than the models and problems of the previous chapter. In Sections 2.4 - 2.6 we consider some other problems where reduced models are of interest.

The usefulness of reduced models derives from the fact that the various efficiency/optimality problems can be treated in a unified manner in the context of these models. We shall consider this treatment in Chapter 3.

2.1. DEFINITIONS.

We start on a somewhat abstract level with the formal definition of a reduced model:

DEFINITION 2.1. A reduced model is a family $\Sigma=(S_t)_{t\varepsilon N}$ where for each $t\varepsilon N$, $S_t \subset R^2$ is closed, $0\varepsilon S_t$, and $S_t+R_+^2 \subset S_t$. ●

Thus, a reduced model is a sequence of closed subsets of the plane, each containing the origin and satisfying a monotonicity property. Note that at present we do not assume any convexity of the sets S_t, $t\varepsilon N$. However, connectedness is a consequence of the property $S_t+R_+^2 \subset S_t$.

For the discussion of efficiency/optimality to follow, the following definitions are needed:

DEFINITION 2.2. Let $\Sigma=(S_t)_{t\varepsilon N}$ be a reduced model. A sequence $(\xi_t)_{t\varepsilon N}$ is an improvement for Σ if

(i) for all $t\varepsilon N$, $(-\xi_t,\xi_{t+1})\varepsilon S_t$
(ii) $\xi_t \geq 0$ and $\xi_t \neq 0$ for some $t\varepsilon N$.

A reduced model Σ is efficient if there is no improvement for Σ. ●

The ideas behind Definition 2 are obviously parallel to those of the efficiency concept for production models and Pareto optimality for consumption models (cf. Chapter 1). But it should be noted that - contrasting with the models in the previous chapter - what is to be efficient is not a single program among several but the whole model as such. This corresponds to the fact - as will be shown by the following sections - that each program in the production or consumption model defines a reduced model of its own.

REMARK 2.3. At this point the reader might wonder whether there is a connection between improvements as defined above and the improving sequences defined in Chapter 1. Now, improving sequences were defined only for weakly efficient/optimal programs, and therefore improvements and improving seqences need not correspond to each other in any direct way. It will be the case, however, as we shall show in the following sections, that the existence of an improving sequence in a production or consumption model implies the existence of an improvement in the corresponding reduced model.

The concepts to follow are of minor importance compared with the previous ones but they are useful for the further development of the theory.

DEFINITION 2.4. A reduced model $\Sigma=(S_t)_{t\in N}$ is <u>weakly efficient</u> if there is no improvement $(\xi_t)_{t\in N}$ for Σ such that $\xi_t=0$ for all but a finite number of indices t. ●

Trivially if Σ is efficient, then Σ is weakly efficient. Certain useful properties of Σ follow from weak efficiency. For convenience, we introduce the notation $\dot{R}^2_+=R^2_+\setminus\{0\}$

THEOREM 2.5. Let $\Sigma=(S_t)_{t\in N}$ be a reduced model and assume that Σ is weakly efficient. Then

(i) for each $t\in N$, $S_t\cap(-R^2_+)=\{0\}$

(ii) if $(\xi_t)_{t\in N}$ is an improvement for Σ then $\xi_t\geq 0$ for all $t\in N$ and there is $T\in N$ such that $\xi_t=0$ for $t<T$ and $\xi_t>0$ for $t\geq T$.

If, moreover, S_t is convex and $S_t+\dot{R}^2_+\subset \mathrm{int} S_t$, each $t\in N$, then there exists a sequence $(p_t)_{t\in N}$ such that for all $t\in N$

(iii) $p_t>0$

(iv) (p_t,p_{t+1}) supports S_t, i.e. $p_tx_1+p_{t+1}x_2 \geq 0$ for all $x=(x_1,x_2)\in S_t$.

<u>Proof</u>: (i). Suppose that $S_\tau\cap(-\dot{R}^2_+)\neq\phi$ for some $\tau\in N$. If $(x_1,x_2)\in S_\tau\cap(-\dot{R}^2_+)$, then $x_1<0$ or $x_2<0$. If $x_1<0$, define $(\xi_t)_{t\in N}$ by $\xi_t=0$, $t\neq\tau$, $\xi_\tau=-x_1$. If $x_1=0$, and hence $x_2<0$, define $(\xi_t)_{t\in N}$ by $\xi_t=0$ for $t\neq\tau+1$ and $\xi_{\tau+1}=x_2$. In either case, we have an improvement for Σ such that $\xi_t=0$ for all but a finite number of indices t, contradicting weak effiency.

(ii). Suppose that $(\xi_t)_{t\in N}$ is an improvement for Σ with $\xi_t<0$, some t, and let $\tau=\min\{t\,|\,\xi_t<0\}$. Then $\tau>1$ and $(-\xi_{\tau-1},\xi_\tau)\in\dot{R}^2_+$, contradicting (i). Thus $\xi_t\geq 0$ for $t\in N$. Moreover, if $\xi_t>0$, then $\xi_{t+1}>0$ by (i). So (ii) holds with $T=\min\{t\,|\,\xi_t>0\}$.

(iii)-(iv). For each t, the convex sets S_t and $-\dot{R}^2_+$ are disjoint, and consequently there is $(p^t_t,p^t_{t+1})\neq 0$ separating the two sets, i.e. $p^t_tx_1+p^t_{t+1}x_2\geq 0$ for $(x_1,x_2)\in S_t$, $p^t_tx_1+p^t_{t+1}x_2<0$ for $(x_1,x_2)\in-\dot{R}^2_+$. This shows that $p^t_t>0$, $p^t_{t+1}>0$ and we may define the seqence $(p_t)_{t\in N}$ inductively by

$p_1=p_1^1$, $p_2=p_2^1$, and for each $t\geq 2$, $p_{t+1}=(p_t/p_t^t)p_{t+1}^t$. Then $(p_t)_{t\epsilon N}$ satisfies (iii)-(iv).●

A sequence $(p_t)_{t\epsilon N}$ with the properties (iii) and (iv) of Theorem 2.5 is called a <u>support</u> for Σ.

2.2. REDUCTION OF THE PRODUCTION MODEL.

Time has come to give examples of reduced models.

Let $(f_t)_{t\epsilon N}$ be a production model (cf. Chapter 1) and $(x_t^o,y_t^o,c_t^o)_{t\epsilon N}$ a feasible program. Define for each $t\epsilon N$ the set $S_t\subset R^2$ by

(1) $$S_t = \{(x-x_t^o,y_{t+1}^o-y)\,|\,y \leq f_t(x)\}$$

We shall be concerned exclusively with $(z_1,z_2)\epsilon S_t$ for which $z_1\leq 0$, $z_2\geq 0$. Each such pair can be interpreted as a feasible deviation in the production plan from (x_t^o,y_t^o). Instead of x_t^o,only $x_t\leq x_t^o$ is used as input. Therefore, output y_{t+1} is smaller than y_{t+1}^o. This is illustrated in Figure 2.1.

We now check that $\Sigma=(S_t)_{t\epsilon N}$, where each S_t is given by (1) is a reduced model: for each $t\epsilon N$, $0\epsilon S_t$ since $y_{t+1}^o\leq f_t(x_t^o)$, and if $u=(u_1,u_2)\epsilon R_+^2$ and $(z_1,z_2)\epsilon S_t$, i.e. $z_1=x_t-x_t^o$, $z_2=y_{t+1}^o-y_{t+1}$ for some (x_t,y_{t+1}) with $y_{t+1}\leq f_t(x_t)$, then also $y_{t+1}-u_2 \leq f_t(x_t+u_1)$ (free disposal), whence

$$(z_1+u_1,z_2+u_2) = ((x_t+u_1)-x_t^o,y_{t+1}^o-(y_{t+1}-u_2)) =$$

$$= (x_t-x_t^o+u_1,y_{t+1}^o-y_{t+1}+u_2)\epsilon S_t.$$

The model constructed above is called the reduced model associated with the production model $(f_t)_{t\epsilon N}$ and the feasible program $(x_t^o,y_t^o,c_t^o)_{t\epsilon N}$. Note that the c_t's are actually unnecessary for the definition of Σ.This should not be surprising in view of the relation $c_t=y_t-x_t$

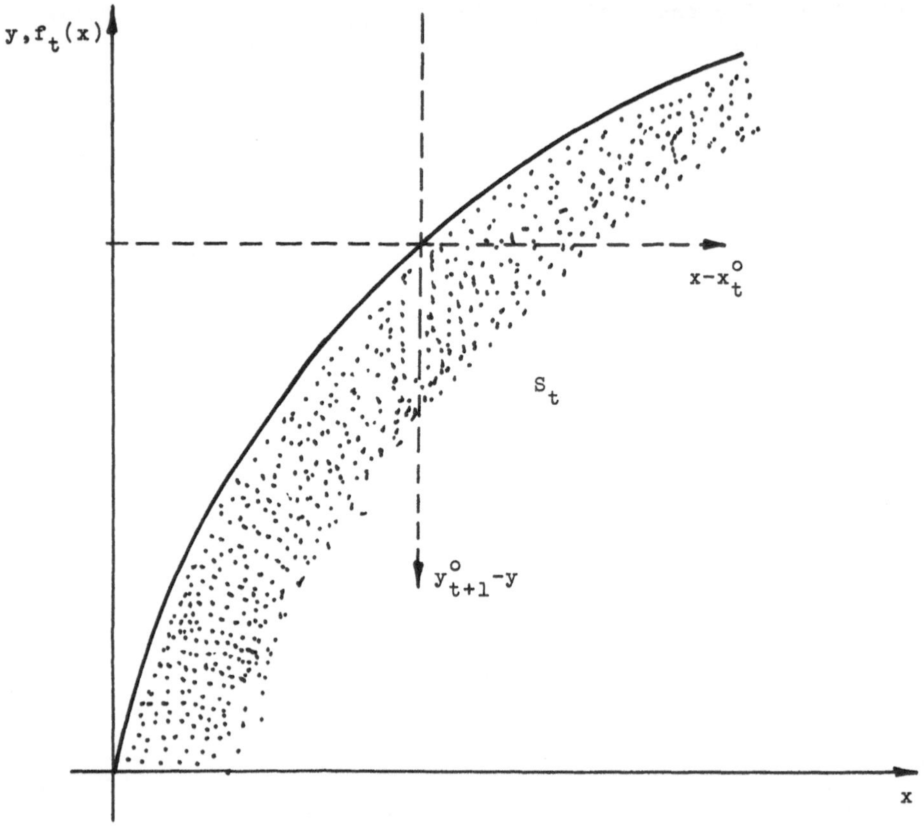

FIGURE 2.1

THEOREM 2.6. Let Σ be the reduced model associated with the production model $(f_t)_{t\varepsilon N}$ and the feasible program $(x_t^o, y_t^o, c_t^o)_{t\varepsilon N}$. Then Σ is efficient if and only if $(x_t^o, y_t^o, c_t^o)_{t\varepsilon N}$ is efficient.

Proof: Let Σ be efficient and suppose that $(x_t', y_t', c_t')_{t\varepsilon N}$ is a feasible program such that $y_1' = y_1^o$, $c_t' \geq c_t^o$ all t and $c_t' > c_t^o$ for some t ; then the sequence $(\xi_t)_{t\varepsilon N}$ with $\xi_t = x_t^o - x_t'$ has the following properties:

(i) for each $t\varepsilon N$, $(-\xi_t, \xi_{t+1}) \varepsilon S_t$

Indeed,

$$x_{t+1}^o - x_{t+1}' = (y_{t+1}^o - c_{t+1}^o) - (y_{t+1}' - c_{t+1}') = y_{t+1}^o - (y_{t+1}' - (c_{t+1}' - c_{t+1}^o))$$

and since $c_{t+1}' - c_{t+1}^o \geq 0$ all t, we have

$$y_{t+1}' - (c_{t+1}' - c_{t+1}^o) \leq y_{t+1}' \leq f_t(x_t')$$

Thus, by monotonicity

$$(x_t'-x_t^o, y_{t+1}^o-y_{t+1}')\varepsilon S_t$$

(ii) $\xi_t = x_t^o - x_t' = (y_t^o-c_t^o)-(y_t'-c_t') \geq c_t'-c_t^o \geq 0.$

If $\xi_t=0$, all t, then $0=y_t^o-(y_t'-(c_t'-c_t^o))$ implies that $y_t'>y_t^o$ for all t with $c_t'>c_t^o$. Let τ be the smallest such t; if $\tau=1$, then $\xi_1>0$. If $\tau>1$, then $y_\tau'>y_\tau^o=f_{\tau-1}(x_{\tau-1}^o)$ implies that $x_{\tau-1}'>x_{\tau-1}^o$, contradicting $\xi_{\tau-1}=0$. Thus $\xi_t\neq0$ for some t.

We conclude that $(\xi_t)_{t\varepsilon N}$ is an improvement for Σ. But Σ was assumed efficient. Thus $(x_t^o,y_t^o,c_t^o)_{t\varepsilon N}$ must be efficient as well.

Conversely, suppose that $(\xi_t)_{t\varepsilon N}$ is an improvement for Σ. Define $(x_t')_{t\varepsilon N}$ by $x_t'=x_t^o-\xi_t$, $t\varepsilon N$. Then, since $(-\xi_t,\xi_{t+1})\varepsilon S_t$ we have that

$$x_{t+1}^o-x_{t+1}' = y_{t+1}^o-y_{t+1}'$$

for some $y_{t+1}'\leq f_t(x_t')$, each $t\varepsilon N$.

Let $\tau=\min\{t|\xi_t\neq0\}$: if $\xi_\tau<0$, then $\tau>1$, $(0,\xi_\tau)\varepsilon S_{\tau-1}$. Thus there is $y_\tau'>y_\tau^o$ for which $y_\tau'\leq f_{\tau-1}(x_{\tau-1}')=f_{\tau-1}(x_{\tau-1}^o)=y_\tau^o$. From this contradiction it follows that $\xi_t>0$.

Define $(y_t'')_{t\varepsilon N}$ by $y_t''=y_t^o$ for $t\leq\tau$, $y_t''=y_t'$ for $t>\tau$, and let $(c_t')_{t\varepsilon N}$ be defined by $c_t'=y_t''-x_t'$, $t\varepsilon N$. Then $c_t'=c_t^o$ for $t\neq\tau$, $c_\tau'>c_\tau^o$, consequently $(x_t^o,y_t^o,c_t^o)_{t\varepsilon N}$ can not be efficient.

2.3. REDUCTION OF THE CONSUMPTION MODEL

We return here to the consumption model with overlapping generations considered in Chapter 1. Recall that in this model we have generations $t=0,1,2,\ldots$, each generation consisting of a finite number of agents (consumers) living for two consecutive time periods. Indicing each consumer by the generation t to which he belongs, as well as his number in this generation, we have an initial generation

$M(0) = \{(\tau,i) | \tau=0\}$

and for each $t \geq 1$, a generation

$M(t) = \{(\tau,i) | \tau=t\}$

In this context, a program is a family $((x^{t,i})_{i\varepsilon M(t)})_{t=0}^{\infty}$ such that $x^{t,i} \varepsilon X^{t,i}$, each (t,i). The condition for feasibility becomes

$$(2) \qquad \Sigma_{i\varepsilon M(t-1)} x_L^{t-1,i} + \Sigma_{i\varepsilon M(t)} x_F^{t,i} = \Sigma_{i\varepsilon M(t-1)} \omega_L^{t-1,i} + \Sigma_{i\varepsilon M(t)} \omega_F^{t,i}$$

for each $t\varepsilon N$, where we use the convention $x^{0,i} = x_L^{0,i}$.

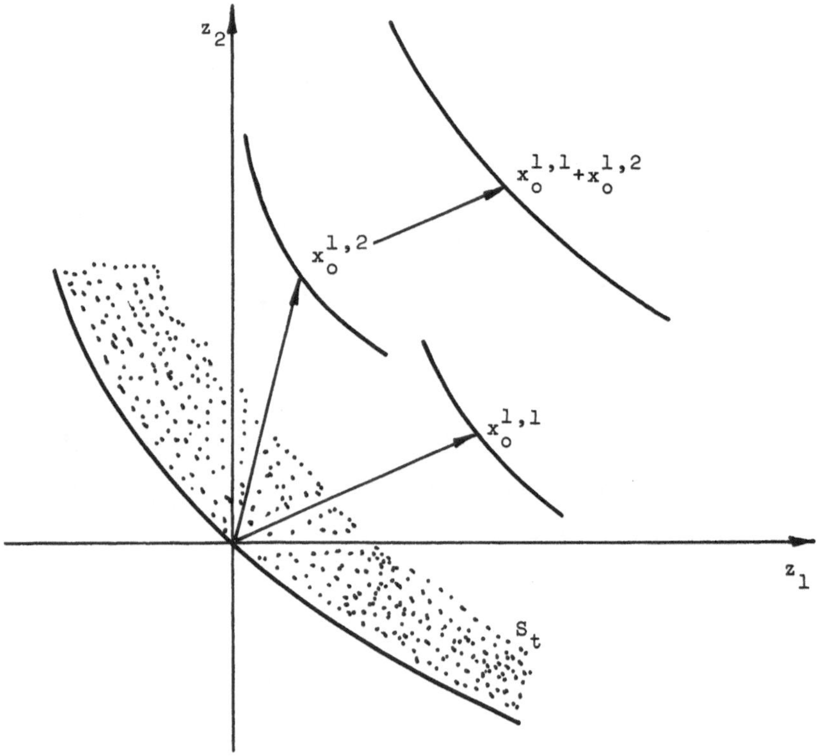

FIGURE 2.2

A program $((x^{t,i})_{i\varepsilon M(t)})_{t=0}^{\infty}$ is Pareto optimal if it is feasible and there is no other feasible program $((\bar{x}^{t,i})_{i\varepsilon M(t)})_{t=0}^{\infty}$ such that for each (t,i), $\bar{x}^{t,i} \varepsilon P^{t,i}(x^{t,i}) \cup \{x^{t,i}\}$ and $\bar{x}^{t,i} \varepsilon P^{t,i}(x^{t,i})$ for some (t,i).

Now we turn to the construction of the reduced model. Let $(x_o^{t,i})_{i\varepsilon M(t)})_{t=0}^{\infty}$ be a feasible program. For $t\varepsilon N$, let

(3) $$S^{t,i} = \{x^{t,i}-x_o^{t,i} \mid x^{t,i}\varepsilon P^{t,i}(x_o^{t,i})\}$$

and let

(4) $$S_t = \text{cl}\,\Sigma_{i\varepsilon M(t)} S^{t,i}$$

In the interpretation the sets $S^{t,i}$ are deviations from the bundle assigned to consumer (t,i) by the initial feasible program, leading to a preferred bundle as shown i Figure 2.2. A similar construction is used in the context of finite economies (cf. Koopmans [1957], Debreu [1959]).

THEOREM 2.7. Let $\Sigma=(S_t)_{t\varepsilon N}$ be a family of sets $S_t \subset R^2$, derived from a feasible program $((x_o^{t,i})_{i\varepsilon M(t)})_{t=0}^{\infty}$ as in (3) and (4). Then

(i) Σ is a reduced model
(ii) Σ is efficient if and only if $((x_o^{t,i})_{i\varepsilon M(t)})_{t=0}^{\infty}$ is Pareto optimal.

Proof: (i). Let $t\geq 1$, $z\varepsilon S_t$ and $u\varepsilon \dot{R}_+^2$. For $\delta>0$ arbitrarily small there are $x^{t,i}\varepsilon P^{t,i}(x_o^{t,i})$ such that

$$\| \Sigma_{i\varepsilon M(t)}(x^{t,i}-x_o^{t,i})-z \| < \frac{\delta}{2}$$

and $u_\delta \varepsilon R_{++}^2$ with $|u_\delta - u| < \frac{\delta}{2}$.

Let m_t be the cardinality of M(t), then, by monotonicity of preferences we have

$$x^{t,i}+\frac{1}{m_t}u_\delta \varepsilon P^{t,i}(x_o^{t,i})+R_{++}^2 \subset P^{t,i}(x_o^{t,i})$$

and moreover,

$$\| \Sigma_{i\varepsilon M(t)}(x^{t,i}+\frac{1}{m_t}u_\delta - x_o^{t,i})-(z+u) \| < \delta$$

We conclude that $z+u\varepsilon S_t$ which shows that S_t is monotone. The remaining properties of S_t are immediate consequences of Assumption 1.3.

(ii). Suppose that $((x_o^{t,i})_{i\varepsilon M(t)})_{t=0}^{\infty}$ is not Pareto optimal and let $((\bar{x}^{t,i})_{i\varepsilon M(t)})_{t=0}^{\infty}$ be a feasible program such that for each (t,i), either $x_o^{t,i}=\bar{x}^{t,i}$ or $\bar{x}^{t,i}\varepsilon P^{t,i}(x_o^{t,i})$; the last relation holding for some t.

Applying (2) twice, we get that

$$(\Sigma_{i\varepsilon M(t-1)}\bar{x}_L^{t-1,i}+\Sigma_{i\varepsilon M(t)}\bar{x}_F^{t,i})-$$

$$-(\Sigma_{i\varepsilon M(t-1)}(x^o)_L^{t-1,i}+\Sigma_{i\varepsilon M(t)}(x^o)_F^{t,i}) = 0$$

or

$$(\Sigma_{i\varepsilon M(t-1)}(\bar{x}_L^{t-1,i}-(x^o)_L^{t-1,i}) = -\Sigma_{i\varepsilon M(t)}(\bar{x}_F^{t,i}-(x^o)_F^{t,i})$$

for each t. Defining $(\xi_t)_{t\varepsilon N}$ by

$$\xi_t = -\Sigma_{i\varepsilon M(t)}(\bar{x}_F^{t,i}-(x_F^o)^{t,i})$$

for $t\varepsilon N$, we have that $(-\xi_t,\xi_{t+1})\varepsilon S_t$ for all t. Clearly $\xi_1\geq 0$ and there is some t such that $\xi_t\neq 0$. We conclude that $(\xi_t)_{t\varepsilon N}$ is an improvement for Σ.

Conversely, let $(\xi_t)_{t\varepsilon N}$ be an improvement for Σ. For any $t\varepsilon N$ with $(-\xi_t,\xi_{t+1})\neq 0$ we have

$$(-\xi_t,\xi_{t+1}) = \lim \Sigma_{i\varepsilon M(t)}s_i^n$$

where $s_i^n\varepsilon S^{t,i}$. Partition $M(t)$ into $\{M^1(t),M^2(t)\}$ where $M^1(t)$ is such that the sequence $(s_i^n)_{n=1}^{\infty}$ is bounded. W.l.o.g. we may assume that $s_i^n\longrightarrow s_i^o$ for each $i\varepsilon M^1(t)$. Let $z^1=\Sigma_{i\varepsilon M(t)}s_i^o$ and $z^2=(-\xi_t,\xi_{t+1})-z^1$.

The set S_t is convex and contains 0. If $S_t\cap(-R_+^2)\neq\{0\}$, then there would be $x_1^{t,i}\varepsilon P^{t,i}(x_o^{t,i})$ such that

$$\Sigma_{i\varepsilon M(t)}x_1^{t,i} < \Sigma_{i\varepsilon M(t)}x_o^{t,i}$$

and $((x_o^{t,i})_{i\varepsilon M(t)})_{t=0}^{\infty}$ could not be Pareto optimal. Thus we may assume that $S_t\cap(-\dot{R}_+^2)=\phi$ and S_t has a support $p=(p_1,p_2)$ with $p_1,p_2>0$, $p_1x_1+p_2x_2\geq 0$, for all $x=(x_1,x_2)\varepsilon S_t$.

Clearly $(ps_i^n)_{n=1}^{\infty}$ must be bounded above for all $i\varepsilon M^1(t)\cup M^2(t)$. It follows that for each $i\varepsilon M^2(t)$,

$$s_i^n/\|s_i^n\|\longrightarrow e(p) \text{ or } s_i^n/\|s_i^n\|\longrightarrow -e(p)$$

where $e(p)$ is a unit vector orthogonal to p. But then $M^2(t)$ partitions into two non-empty subsets $M^{2,1}(t)$, $M^{2,2}(t)$, where

$$\{x\varepsilon R^2 \mid px \geq 0,\ x_1 \geq 0\} \subset clS^{t,i},\ i\varepsilon M^{2,1}(t)$$

$$\{x\varepsilon R^2 \mid px \geq 0,\ x_2 \geq 0\} \subset clS^{t,i},\ i\varepsilon M^{2,2}(t)$$

thus $\Sigma_{i\varepsilon M^2(t)} clS^{t,i} = \{x\varepsilon R^2 \mid px \geq 0\}$.

We conclude that

$$z^2 \varepsilon \Sigma_{i\varepsilon M^2(t)} clS^{t,i}$$

consequently $(-\xi_t, \xi_{t+1}) \varepsilon \Sigma_{i\varepsilon M(t)} clS^{t,i}$. The remaining part of the proof is straightforward and is left to the reader.

REMARK 2.8. A crucial point in the proof above was that $cl\Sigma_i C_i \subset \Sigma_i clC_i$ for certain subsets $C_1,\dots,C_m$ of R^2. Here the dimension played an important role. The counterpart of Theorem 2.7 (ii) does not hold for dimensions higher than 2 without additional assumptions (cf. part II).

2.4.* A DOUBLE INFINITY MODEL OF PRODUCTION

The following model is an example of an application of reduced models, where time goes backwards rather than forwards. It is mentioned in order to show that the reduced model construction can be applied in other situations than the production and consumption models treated in Chapter 1. Otherwise, the present model is of limited interest.

Suppose that for each $t\varepsilon Z$ (the set of integers) we are given a production function $f_t: R_+ \longrightarrow R_+$ satisfying Assumption 1.1. A program, now, is a doubly infinite sequence $(x_t)_{t\varepsilon Z}$. The program $(x_t)_{t\varepsilon Z}$ is said to be feasible if for each $t\varepsilon Z$, $x_{t+1} \leq f_t(x_t)$. The production model $(f_t)_{t\varepsilon Z}$ is said to be productive if there exists a feasible program $(x_t)_{t\varepsilon Z}$ with $x_t > 0$, for all $t\varepsilon Z$.

Note that, for arbitrary $\tau\varepsilon Z$, if there is a feasible program $(x_t)_{t\varepsilon Z}$ with $x_\tau > 0$, then $x_t > 0$ for $t < \tau$ since by (i) $f_t(0)=0$, each t. Also there is a feasible program $(x_t')_{t\varepsilon Z}$ with $x_t' = x_t$ for $t \leq \tau$ and $x_t' > 0$ for $t > \tau$: let $x_{\tau+k} = f_{\tau+k}(\dots f_\tau(x_\tau)\dots)$. Thus the problem of productiveness amounts to whether or not some positive quantity of goods can be produced in some time period t which without loss of generality may assumed to be less than 0.

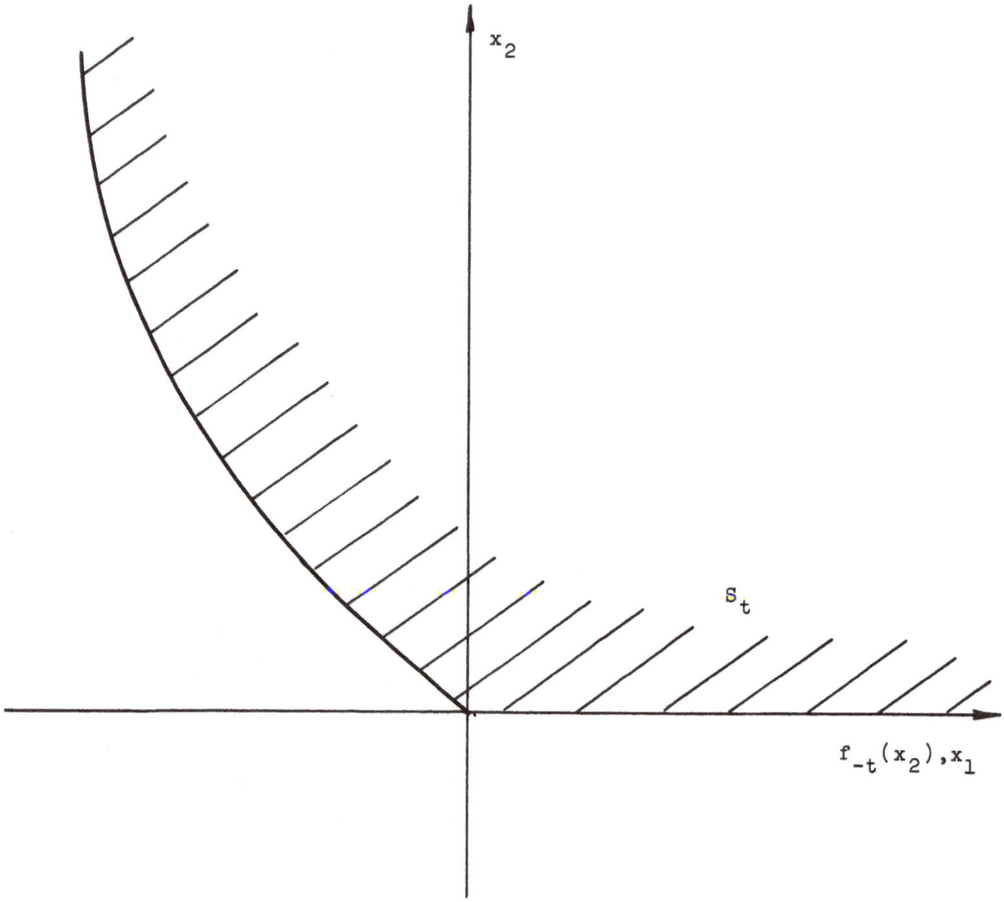

FIGURE 2.3

Define $\Sigma=(S_t)_{t\in N}$ by

$$S_t = \{(x_1,x_2) \mid -x_1 \leq f_{-t}(x_2),\ x_2 \geq 0\}$$

Then for each $t\in N$, $0\in S_t$ and S_t is closed. Moreover if $x\in S_t$ and $u\in\dot{R}^2_+$, then

$$-(x_1+u_1) = -x_1-u_1 \leq f_{-t}(x_2) \leq f_{-t}(x_2+u_2)$$

where the last inequality follows from the fact that f_{-t} is non-decreasing. Thus $x+u\in S_t$ and Σ is a reduced model.

Suppose that Σ is not efficient, and let $(\xi_t)_{t\epsilon N}$ be an improvement for Σ. Let $\tau=\min\{t|\xi_t\neq 0\}$. Since by the construction of S_t we have $S_t\subset\{(x_1,x_2)|x_2\geq 0\}$, we get from $(-\xi_{t-1},\xi_t)\epsilon S_{t-1}$ that $\xi_t>0$ can be produced at time $\tau\leq 0$, thus the model $(f_t)_{t\epsilon Z}$ is productive. Conversely, a feasible program $(x_t)_{t\epsilon Z}$ with $x_t>0$, all t,defines an improvement $(\xi_t)_{t\epsilon N}$ for Σ with $\xi_t=x_{-t}$, all $t\epsilon N$.

Thus, the production model is productive if and only if the corresponding reduced model is <u>not</u> efficient.

2.5.* VARIATIONS OVER THE THEME: DISCOUNTED UTILITY MAXIMIZATION.

In the models considered in Chapter 1 and again in the present chapter, Sections 2 and 3, we discussed efficiency problems; here the dates (or generations) are given equal weight no matter how distant in the future they may be.

This type of models with no discounting is the principal topic of this book, and we shall depart from it only occasionally to indicate the wider applicability of the general method.

An alternative to the equal treatment with respect to dates or generations is the introduction of explicit weighting of consumption at different dates. We shall illustrate this kind of discounted utility models by an example in the context of the production model of Section 1.1 or 2.2.

Thus, let $(f_t)_{t\epsilon N}$ be a production model. A program $(x_t,y_t,c_t)_{t\epsilon N}$ is said to be <u>admissible</u> if it is feasible and $\Sigma_{t=1}^{\infty}c_t$ is a convergent series, to be written $\Sigma_{t=1}^{\infty}c_t<\infty$, and <u>optimal</u> if it is admissible and there is no other admissible program $(x_t',y_t',c_t')_{t\epsilon N}$ such that $y_1'=y_1$ and $\Sigma_{t=1}^{\infty}c_t'>$ $>\Sigma_{t=1}^{\infty}c_t$.

The sum of consumption streams to be maximized may seem a rather special criterion. However, this is equivalent to maximizing

$$\Sigma_{t=1}^{\infty} d_t c_t, \quad d_t = \Pi_{\tau=1}^{t}(1+\rho_\tau)$$

where ρ_τ is an interest rate (from τ-1 to τ), since one problem can be obtained from the other by changing units at each date t. The more general criterion $\Sigma_{t=1}^{\infty} u_t(c_t)$ could also be treated; however, results for this case may be more easily obtained in the context of the many-goods models, but we shall not pursue this matter further.

Note that due to the admissibility condition the relation between efficient and optimal programs is not altogether trivial (in contrast with the situation in finite-dimensional models). At present, we shall make a simplifying assumption so as to avoid problems with convergence of $\Sigma_{t=1}^{\infty} c_t$.

ASSUMPTION 2.9. The production model $(f_t)_{t \in N}$ satisfies

(1) Differentiability: $f_t \in C^1(R_+)$, each t, and there is R>0 such that $|f_t'|<R$, all t.

(2) Strong Uniform Boundedness: There are numbers $(r_t)_{t \in N}$ such that $r_t>0$, all t, $\Sigma_{t=1}^{\infty} r_t < \infty$, and for every feasible program $(x_t, y_t, c_t)_{t \in N}$, there is T such that $y_t < r_t$ for $t \geq T$.

REMARK 2.10. The restrictiveness of the above assumption is connected with the fact that the y's rather than $c_t = y_t - x_t$ are assumed to be bounded by $(r_t)_{t \in N}$. The latter assumption is usually made in order to prove existence of optimal programs.

THEOREM 2.11. Let $(f_t)_{t \in N}$ be a production model satisfying Assumptions 1.1 and 2.9, and let $(x_t^o, y_t^o, c_t^o)_{t \in N}$ be a feasible program. Then $(x_t^o, y_t^o, c_t^o)_{t \in N}$ is optimal if and only if there is no feasible program $(x_t', y_t', c_t')_{t \in N}$ such that $y_1' = y_1^o$ and

(5) $$0 < \Sigma_{t=1}^{\infty}(f_t'(x_t^o)-1)(x_t'-x_t^o) < \infty.$$

Proof: Suppose that $(x_t^o, y_t^o, c_t^o)_{t \in N}$ is not optimal, and let $(x_t', y_t', c_t')_{t \in N}$ be an admissible program with $\Sigma_{t=1}^{\infty} c_t' > \Sigma_{t=1}^{\infty} c_t^o$, or $\Sigma_{t=1}^{\infty}(c_t'-c_t^o)>0$. Then $c_1'-c_1^o = -(x_1'-x_1^o)$, and, by concavity of f_t, all t,

$$c_t'-c_t^o = y_t'-x_t'-(y_t^o-x_t^o) = y_t'-y_t^o-(x_t'-x_t^o) \leq$$
$$\leq f_{t-1}'(x_{t-1}^o)(x_{t-1}'-x_{t-1}^o)-(x_t'-x_t^o)$$

for $t \leq 2$. Thus, for $T \geq 2$ we have

$$\Sigma_{t=1}^{T}(c_t'-c_t^o) \leq \Sigma_{t=1}^{T-1}(f_t'(x_t^o)-1)(x_t'-x_t^o)-(x_T'-x_T^o).$$

Since the right hand side is less than $2(R+1)\Sigma_{t=1}^{T-1}r_t+2r_t$ we get by letting $T \to \infty$ that

$$0 < \Sigma_{t=1}^{\infty}(c_t'-c_t^o) \leq \Sigma_{t=1}^{\infty}(f_t'(x_t^o)-1)(x_t'-x_t^o)<\infty$$

which is (5).

Conversely, suppose that $(x_t',y_t',c_t')_{t\varepsilon N}$ is a feasible program satisfying (5). By Assumption 2.9, we have

$$|x_t'-x_t^o| \leq |y_t'|+|y_t^o| \leq 2r_t$$

for all t, hence $\Sigma_{t=1}^{\infty}(x_t'-x_t^o)<\infty$. Also, by the same argument, $\Sigma_{t=1}^{\infty}(c_t'-c_t^o)<\infty$.

Choose T large enough so that

$$|\Sigma_{t=T+1}^{\infty}(c_t'-c_t^o)| < \frac{r}{2}$$
$$\Sigma_{t=T}^{\infty}(f_t'(x_t^o)-1)(x_t'-x_t^o) + (x_T'-x_T^o) < \frac{r}{2}$$

where $r=\Sigma_{t=1}^{\infty}(f_t'(x_t^o)-1)(x_t'-x_t^o)$. For each $\lambda\varepsilon[0,1]$, the program $(x_t^\lambda,y_t^\lambda,c_t^\lambda)_{t\varepsilon N}$, where

$$x_t^\lambda = \lambda x_t+(1-\lambda)x_t^o$$
$$y_t^\lambda = \lambda y_t+(1-\lambda)y_t^o$$
$$c_t^\lambda = \lambda c_t+(1-\lambda)c_t^o$$

is feasible, and for $\lambda>0$ small enough we have

$$\frac{\lambda r}{2} \leq \frac{1}{2} \quad \Sigma_{t=1}^{T-1}(f_t'(x_t^o)-1)(x_t^\lambda-x_t^o)-(x_T^\lambda-x_T^o) \leq \Sigma_{t=1}^{T}(c_t^\lambda-c_t^o)$$

Also, $|\Sigma_{t=T+1}^{\infty}(c_t^\lambda-c_t^o)|<\frac{\lambda r}{2}$, so

$$\Sigma_{t=1}^{\infty}(c_t^\lambda-c_t^o) > \lambda(\frac{r}{2} - \frac{r}{2}) = 0$$

It follows that $(x_t^o,y_t^o,c_t^o)_{t\varepsilon N}$ cannot be optimal.●

COROLLARY. Let $(x_t^o, y_t^o, c_t^o)_{t\varepsilon N}$ be a feasible program, and define for each $t\varepsilon N$,

$$S_t = \{(z_t, z_{t+1}) | z_t = (f_t'(x_t^o)-1)(x_t^o - x_t),$$
$$z_{t+1} = (f_{t+1}'(x_{t+1}^o)-1)(y_{t+1}-y_{t+1}^o),\ y_{t+1} \leq f_t(x_t)\}.$$

Then $\Sigma=(S_t)_{t\varepsilon N}$ is a reduced model, and $(x_t^o, y_t^o, c_t^o)_{t\varepsilon N}$ is optimal if and only if there is a sequence $(\xi_t)_{t\varepsilon N}$ such that

(i) $(-\xi_t, \xi_{t+1})\varepsilon S_t$ for all t

(ii) $\Sigma_{t=1}^{\infty}\xi_t > 0$.●

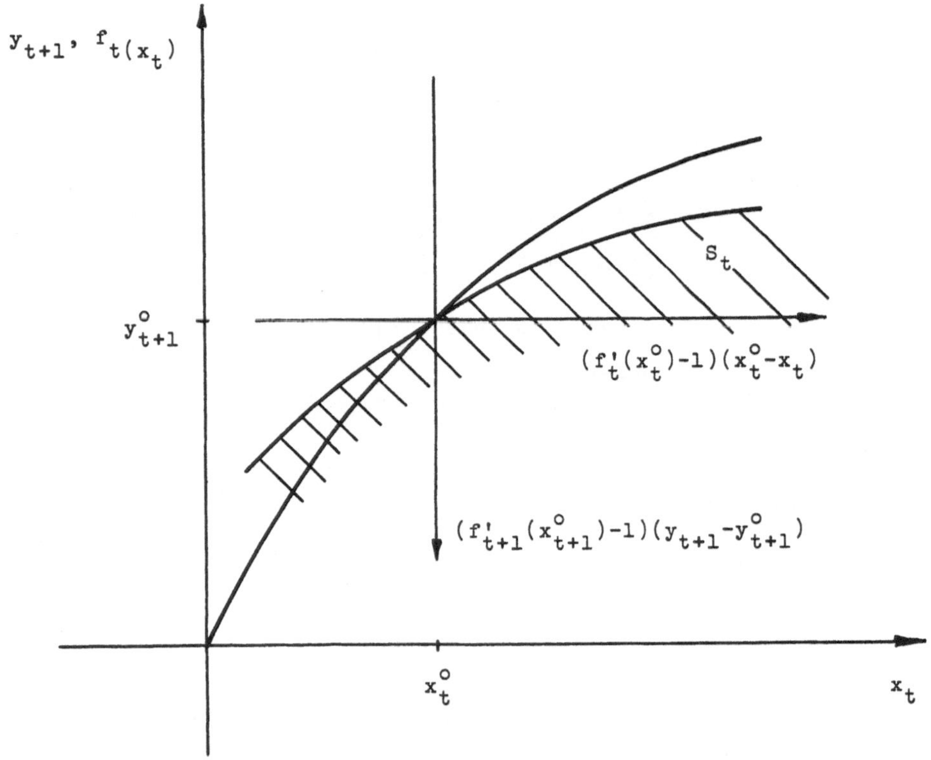

FIGURE 2.4

Note that apart from a rescaling with factors $(f_t'(x_t^o)-1)$ and $(f_{t+1}'(x_{t+1}^o)-1)$, the construction of the reduced model is the same as in Section 2.2. The notion of an improvement, however, is different. It will be seen later that the reduced model captures the feasible set of an optimization problem, whereas the criterion is taken care of by the notion of an improvement.

2.6.* FURTHER APPLICATIONS OF REDUCED MODELS: THE PURE BIRTH PROCESS.

In our definition (Definiton 2.1) of the reduced model we demanded that each S_t should satisfy the monotonicity property $S_t+R^2_+\subset S_t$. In the applications considered hitherto, this monotonicity was an immediate consequence of other, economically reasonable, assumptions as e.g. free disposal. Furthermore, we shall argue at a later point that this monotonicity property is rather natural for reduced models.

However, from the formal point of view, our theory makes perfect sense also without this assumption, and below we consider an application of such (weak) reduced models.

DEFINITION 2.12. A <u>weak reduced model</u> is a family $\Sigma=(S_t)_{t\epsilon N}$ where for each $t\epsilon N$, $S_t\subset R^2$ is closed, and $0\epsilon S_t$. ●

We consider a weak reduced model arising from a problem in the theory of stochastic processes. Our main interest is not in the particular problem, which anyway is easily solved using other methods; rather we want to emphasize that the method of reduced models can be used in a broad variety of contexts.

Following Feller ([1968],p.402) we define a <u>pure birth process</u> as a family $P=(P_n(\cdot))_{n\epsilon N}$ of functions $P_n:R_+\to[0,1]$ satisfying the system of differential equations

(6) $$P_n'(t)=-\lambda_nP_n(t)+\lambda_{n-1}P_{n-1}(t) \text{ for } n>1$$

(7) $$P_1'(t)=-\lambda_1P_1(t)$$

for some family $(\lambda_n)_{n\epsilon N}$ of positive real numbers, together with the initial conditions $P_1(0)=1$, $P_n(0)=0$ for $n>1$.

In the interpretation, we have a system which can be in states E_n, $n=1,2,\ldots$. At time t, the probability that the system is in state E_n is $P_n(t)$. It is assumed that if the system is in E_n at time t, then the probability that a change to E_{n+1} takes place in the small time interval $(t,t+h)$ is $\lambda_nh+o(h)$. Then

$$P_n(t+h) = P_n(t)(1-\lambda_n h)+P_{n-1}(t)\lambda_{n-1}h+o(h),$$

and (6)-(7) follow for $h \to 0$.

It can be shown that $\Sigma_{n=1}^{\infty} P_n(t) \leq 1$ for all $t \in R_+$, and we shall be interested in whether or not equality obtains. The number $1-\Sigma_{n=1}^{\infty} P_n(t)$ may be interpreted as the probability that the system has "exploded" at time t and hence is in none of the states $(E_n)_{n\in N}$.

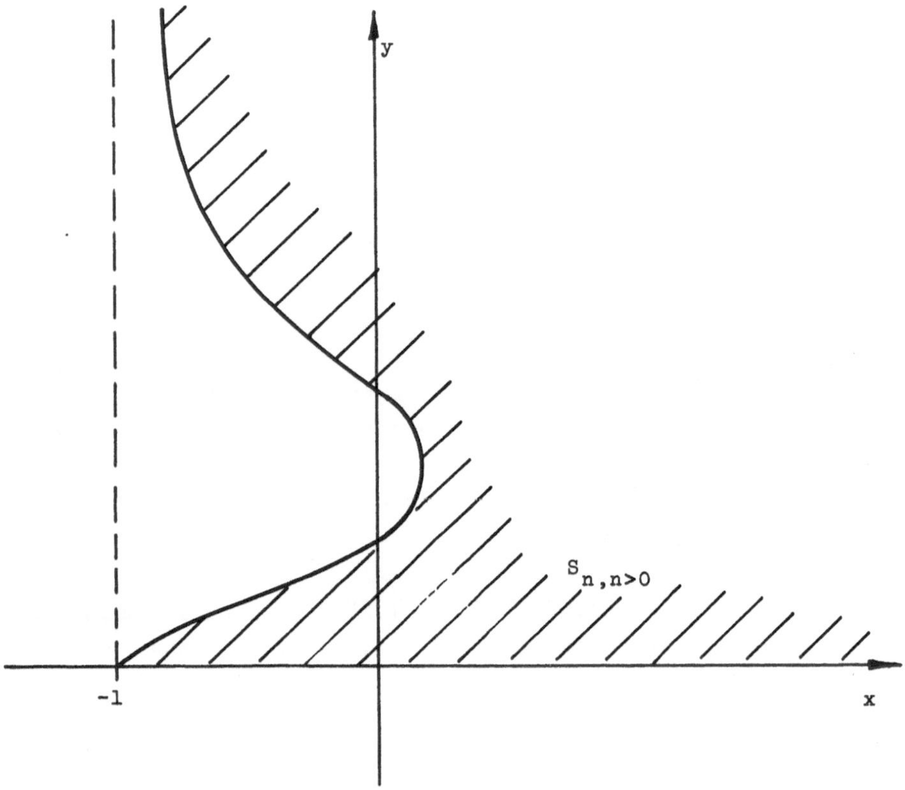

FIGURE 2.5

The birth process $P=(P_n)_{n\in N}$ is said to be <u>divergent</u> if for some $t>0$, $\Sigma_{n=1}^{\infty} P_n(t)<1$. Define a weak reduced model $\Sigma=(S_n)_{n\in N}$ by

$$S_n = \{(x,y) \mid y \geq 0, x \geq P_n(y)-1\}$$

Further, we define an <u>improvement</u> for Σ to be a sequence $(\xi_n)_{n\in N}$ such that

(i) there exists $t > 0$ such that $(-\xi_n, t) \in S_n$, all $n \in N$,

(ii) $\Sigma_{n=1}^{\infty}(\xi_n + 1) < 1$

We leave it to the reader to check that $\Sigma_{n=1}^{\infty} P_n(t) = 1$ for all t if and only if there is no improvement for Σ, or, equivalently, P is divergent if and only if there is an improvement for the (weak) reduced model Σ corresponding to P. We return to the birth process in Chapter 4.

CHAPTER 3: EFFICIENCY CRITERIA

In this chapter we investigate conditions under which a reduced model Σ is efficient. We establish a general efficiency condition which is valid for a large class of reduced models.

Due to its abstract character, the general efficiency criterion may be difficult to use in many applications. Therefore we consider parametrized versions of the criterion and show that straightforward applications of the general conditions will yield several efficiency criteria known from the literature as well as some new ones.

The chapter closes with some considerations of the applicability of the methods developed to the related model of discounted utility maximization (cf. Section 2.5).

3.1. REDUCED MODELS AND COMPOSITION OF RELATIONS.

In this section we take a closer look at the sets S_t in the reduced model $\Sigma=(S_t)_{t\in N}$. In our characterization of efficient reduced models we shall need some properties of the sets S_t apart from those stated in Definiton 2.1.

Let Ω be the family of all closed subsets S of R^2 satisfying

(Ω1) $0 \in S$

(Ω2) $S + R^2_+ \subset S$

(Ω3) $S \cap (-R^2_+) = \{0\}$

(Ω4) S is star-shaped with respect to 0, i.e. if $x \in S$, $\lambda \in [0,1]$, then $\lambda x \in S$.

We shall consider reduced models $\Sigma = (S_t)_{t \in N}$ such that each S_t belongs to Ω. By Theorem 2.1, (Ω3) is fulfilled for each S_t whenever Σ is weakly efficient (recall that in this case improvements have non-negative components). The assumption that each S_t is star-shaped is, of course, a restriction, but it is weaker than convexity and could in fact be further weakened.

A certain composition operation on Ω turns out to be useful:

LEMMA 3.1. Let $S, S' \in \Omega$ and let

$$SoS' = \{(x,y) \in R^2 \mid \exists z \in R : (x,z) \in S, (-z,y) \in S'\}.$$

Then $SoS' \in \Omega$.

<u>Proof</u>: First of all, we show that SoS' is closed. Let $((x^n,y^n))_{n \in N}$ be a sequence converging to (x^o,y^o) such that $(x^n,y^n) \in SoS'$ for each n. Thus, there are $z^n \in R$ such that $(x^n,z^n) \in S'$, $(-z^n,y^n) \in S'$, each n. If the sequence $(z^n)_{n \in N}$ is bounded, then it has an accumulation point z^o. It follows that $(x^o,z^o) \in S$, $(-z^o,y^o) \in S'$, i.e. $(x^o,y^o) \in SoS'$. If $(z^n)_{n \in N}$ is unbounded, let $\lambda^n = \min\{1, |z^n|^{-1}\}$. Then $(\lambda^n x^n, \lambda^n z^n) \in S$, $(-\lambda^n z^n, \lambda^n y^n) \in S'$, all n, so either $(0,-1) \in S$ or $(-1,0) \in S'$. In either case we have a contradiction.

It remains to check that SoS' satisfies (Ω1)-(Ω4):(Ω1) is obvious. (Ω2): If $(x,y) \in SoS'$ there is z such that $(x,z) \in S$, $(-z,y) \in S'$. For $u \in R^2$ we have $(x_1+u_1,z) \in S$, $(-z,y+u_2) \in S'$ whence $(x+u_1,y+u_2) \in SoS'$. (Ω3): If $(x,y) \in SoS' \cap (-R^2_+)$, then for some $z \in R$ $(x,z) \in S$, $(-z,y) \in S'$, which implies that $(x,z) \in -R^2_+$ or $(-z,y) \in -R^2_+$. We conclude that $(x,z) = 0$. (Ω4): If $(x,z) \in S$, $(-z,y) \in S'$ and $\lambda \in [0,1]$, then $(\lambda x, \lambda z) \in S$, $(-\lambda z, \lambda y) \in S'$, so $(\lambda x, \lambda y) \in SoS'$.●

Note that the notation SoS' is in accordance with usual notation for composition of relations. It turns out to be convenient to write the composition from left to right rather than as usual from right to left.

Recall that a semigroup (A,o) is a set A together with an associative composition rule o on A (i.e. $o: A\times A \to A$ satisfies $(a_1 o a_2) o a_3 = a_1 o (a_2 o a_3)$ for all $a_1, a_2, a_3 \varepsilon A$).

THEOREM 3.2. (Ω, o) is a semigroup.

Proof: This is a consequence of Lemma 3.1 and

$$(SoS')oS" = \{(x,y) \mid \exists z_1, z_2 : (x,z_1)\varepsilon S, (-z_1,z_2)\varepsilon S', (-z_2,y)\varepsilon S"\} = So(S'oS"). \bullet$$

The composition o can be given an interpretation in the context of a given reduced model $\Sigma = (S_t)_{t\varepsilon N}$. For each t, one may think about S_t as consisting of various possible trade-offs between dates t and t+1. Thus, in the production model, if $(z_t, z_{t+1})\varepsilon S_t$ and $z_t < 0$, then $|z_t|$ more units of the good can be consumed at date t against a reduction in next date's output of z_{t+1}. In the consumption model, elements of the reduced model have a similar interpretation. Now, if $(x,y)\varepsilon S_t o \dots o S_T$ for $T \geq t$, then this interpretation gives that $|x|$ additional units can be had at date t against some decrease in production or compensation y at time T+1. Thus the sets $S_t o \dots S_T$ for $T \geq t$ give the possibilities of substitution between date t and the future.

3.2. EXAMPLES

In this section we illustrate the composition o introduced in Section 3.1 in some simple cases and show how it relates to our main theme, efficiency.

EXAMPLE 3.3. Halfspaces. For $(p_1,p_2)\varepsilon R^2_{++}$, let

$$H(p_1,p_2) = \{(x_1,x_2) \mid p_1x_1 + p_2x_2 \geq 0\}$$

Clearly, $H(p_1,p_2)\varepsilon\Omega$. Let $(p_1',p_2')\varepsilon R^2_{++}$; we want to determine $H = H(p_1,p_2)oH(p_1',p_2')$. Without loss of generality we may assume that prices are normalized so that $p_2 = p_1'$.

Let $(x,y)\varepsilon H$. Then there is $z\varepsilon R$ so that

$$p_1x+p_2z \geq 0$$
$$-p_1'z+p_2'y \geq 0$$

which implies $p_1x+p_2'y\geq 0$. On the other hand, if x and y are such that $p_1x+p_2'y\geq 0$, choose z so that $p_2'z=p_2'y$. Then the two relations above are satisfied, hence $(x,y)\varepsilon H$. Thus we have shown that

$$H(p_1,p_2)oH(p_1',p_2')=H(p_1,\frac{p_2'}{p_1'}\ p_2).\bullet$$

As a special case, we note that $H(1,1)oH(1,1)=H(1,1)$, thus $H(1,1)$ is idempotent. The above example can be exploited to give the following useful lemma:

LEMMA 3.4. Let $S_1,S_2\varepsilon\Omega$, $S_1\subset H(p_1,p_2)$, $S_2\subset H(p_1',p_2')$. Then

$$S_1oS_2\subset H(p_1,\frac{p_2'}{p_1'}\ p_2)$$

<u>Proof</u>: Immediate from Example 3.3 and the fact that the composition o respects the inclusion ordering.●

EXAMPLE 3.5. <u>Subsemigroups of Ω</u>. A subset Ω' of Ω such that $S,S'\varepsilon\Omega'$ implies $SoS'\varepsilon\Omega'$ is called a <u>subsemigroup</u>. By Lemma 3.4, the set $\Omega^1=\{S\varepsilon\Omega|S\subset H(1,1)\}$ is a subsemigroup.

Let $S\varepsilon\Omega^1$ and $(x,y)\varepsilon S$. Since $(x,-x)$ and $(-y,y)$ belong to $H(1,1)$, we have $SoH(1,1)=S$, $H(1,1)oS=S$. Thus $H(1,1)$ is neutral for the composition ($H(1,1)$ is a unit in the semigroup Ω^1). Note that Ω^1 also has a zero element, since $SoR_+^2=R_+^2oS=R_+^2$ for all $S\varepsilon\Omega^1$.

In the following two examples we consider reduced models $\Sigma=(S_t)_{t\varepsilon N}$ such that $S_t\varepsilon\Omega^1$, all $t\varepsilon N$. Since any such Σ has $(1,1,...)$ as support they are somewhat more transparent than general reduced models. Moreover, as will become evident in later sections, results for such models extend in a simple way to the general case. ●

EXAMPLE 3.6. <u>The Malinvaud Model</u>. Let Σ be a reduced model such that for each t, S_t belongs to the class

$$\Omega^2=\{S\varepsilon\Omega^1|S=H(1,1)\cap\{(x,y)|x\geq -b_S\},b_S>0\}\cup\{H(1,1)\}.$$

Note that Ω^2 is a subsemigroup of Ω^1: If $S,S'\varepsilon\Omega^2$, then

$$S \circ S' = \{(x,y) \mid x+y \geq 0, x \geq -\min\{b_S, b_{S'}\}\}.$$

Indeed, if $(x,z) \in S$, then $x \geq -b_S$ and $x \geq -z$; if also $(-z,y) \in S'$, then $-z \geq -b_{S'}$, thus $x \geq -b_{S'}$ whence $x \geq -\min\{b_S, b_{S'}\}$; the converse is straightforward. Writing $\min\{a,\infty\} = \min\{\infty,a\} = a$ for all $a \in R_+$, we see that the relation above holds for all $S, S' \in \Omega^2$. The reader may notice that the map $S \to b_S$ is a semigroup homomorphism from $(\Omega^2, \circ)$ to (R^2, ∇), where $a \nabla b$ stands for $\min\{a,b\}$. ●

An element of Ω^2 looks as illustrated in Figure 3.1.

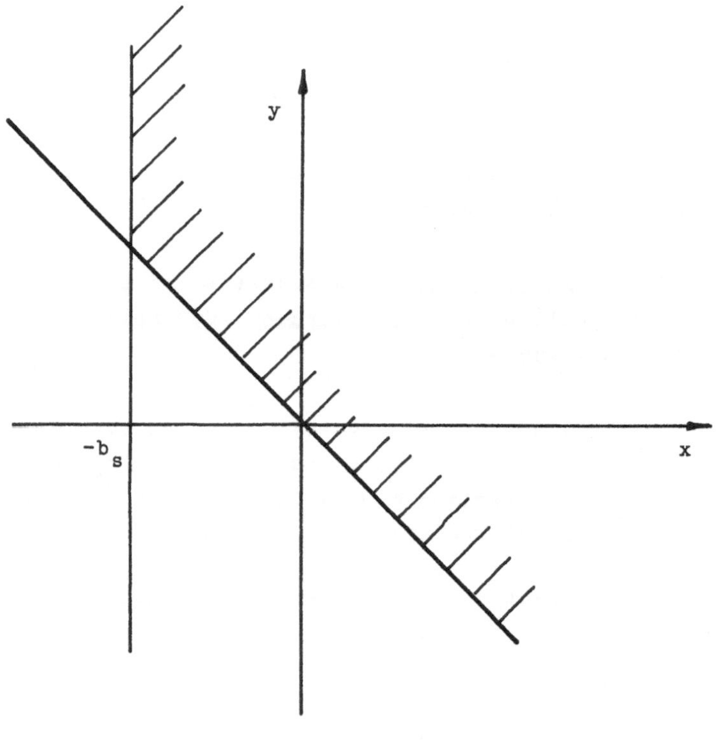

FIGURE 3.1

Returning to the reduced model Σ, suppose that Σ is not efficient, thus there is an improvement $(\xi_t)_{t \in N}$ for Σ. For simplicity, we assume that $\xi_1 > 0$.

By the definition of an improvement,

$(-\xi_1,\xi_2)\varepsilon S_1$
$(-\xi_2,\xi_3)\varepsilon S_2$
.
.
.
$(-\xi_t,\xi_{t+1})\varepsilon S_t$

It follows that $(-\xi_1,\xi_{t+1})\varepsilon S_1 o\ldots oS_t$, each t. Applying our results on composition in Ω^2, we get that $\xi_1\leq\min\{b_1,\ldots,b_t\}$, where $b_t=b_{S_t}$, all $t\varepsilon N$.

Now, turning the problem around we may conclude that if $\inf\{b_t|t\varepsilon N\}=0$, then there can be no improvements $(\xi_t)_{t\varepsilon N}$ for Σ with $\xi_1>0$. Then, clearly, if

$$\liminf_{t\to\infty}b_t=0$$

we have that there can be no improvements for Σ with $\xi_t>0$, some t, so Σ must be efficient.

Thus we have developed an efficiency criterion for this type of models by considerations involving the composition o. The "liminf"-criterion was introduced originally by Malinvaud [1953].

EXAMPLE 3.7. <u>The Hyperbola Subsemigroup</u>. Let Ω^3 be the subset of Ω^1 consisting of all sets

$$S^a=\{(x,y)|x+y\geq0,x+y+axy\geq0\}$$

for $a\varepsilon R_+$, together with the set $S^\infty=R_+^2$.

Again, Ω^3 is a subsemigroup of Ω^1. We show that $S^a oS^b=S^{a+b}$ for all $a,b\varepsilon R_+$: Let $(x,z)\varepsilon S^a$, $(-z,y)\varepsilon S^b$. Then

(1) $x+z+axz \geq 0$

(2) $-z+y-bzy \geq 0$.

Suppose that $x<0$. Then $z\geq0$, $-z\leq0$, whence $y\geq0$. From (2) we get that $z\leq y(1+by)^{-1}$, which inserted in (1) yields $x+y+(a+b)xy\geq0$. If $x\geq0$ we get the same expression by evaluating z from (1) and inserting in (2). Thus $(x,y)\varepsilon S^{a+b}$.

Conversely, if $(x,y)\varepsilon S^{a+b}$ and $y\geq0$, then $(-y(1+by)^{-1},y)\varepsilon S^b$, and

$$x+\frac{y}{1+by}+\frac{axy}{1+by}=(1+by)^{-1}(x+y+bxy+axy)\geq 0$$

so $(x,y(1+by)^{-1})\in S^a$.

If $y<0$, then $x\geq 0$ and $(x,-x(1+ax)^{-1})\in S^a$, $(x(1+ax)^{-1},y)\in S^b$ by a similar argument. Thus $S^a o S^b=S^{a+b}$.

An element of Ω^3 is pictured in Figure 3.2.

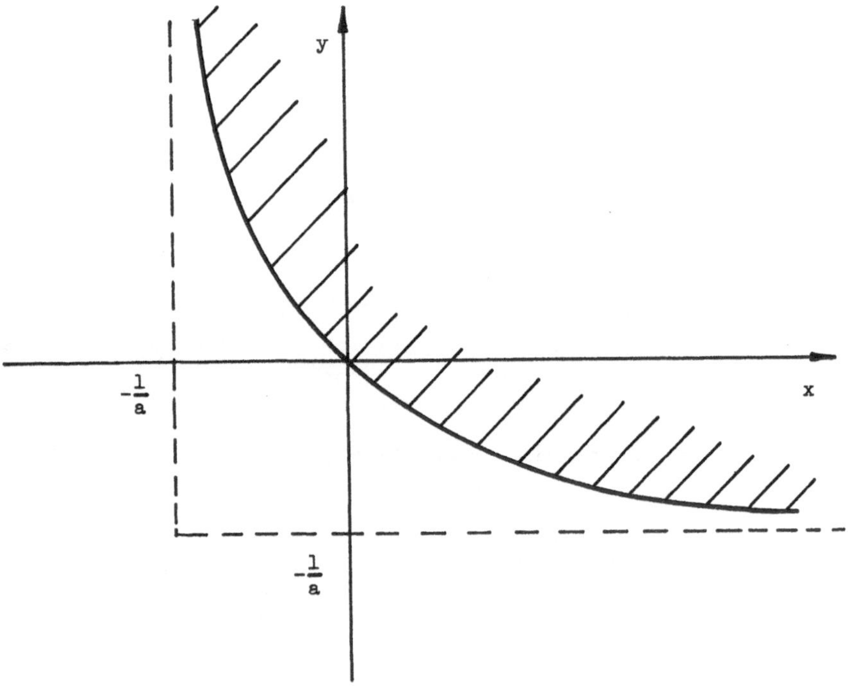

FIGURE 3.2

Note that S^a is symmetric around the diagonal (if $(x,y)\in S^a$ then $(y,x)\in S^a$). Also, if $(x,y)\in S^a$ and $x<0$, then $y\geq 0$, consequently

$$x\geq -y(1+ay)^{-1}=-((1/y)+a)^{-1}\geq -a^{-1}.$$

Now, consider a reduced model $\Sigma=(S_t)_{t\in N}$ such that $S_t\in\Omega^3$, each $t\in N$. As in the previous example, we have that if $(\xi_t)_{t\in N}$ is an improvement with $\xi_1>0$, then

$$(-\xi_1,\xi_2)\epsilon S_1=S^{a_1},$$

$$(-\xi_1,\xi_3)\epsilon S_1 o S_2=S^{a_1+a_2}$$

$$\vdots$$

$$(-\xi_1,\xi_{t+1})\epsilon S_1 o\ldots o S_t=S^{a_1+\ldots+a_t}$$

But in this case, $\xi_1\leq(a_1+\ldots+a_t)^{-1}$ for all t, which again means that $\Sigma_{t=1}^{\infty}a_t\leq 1/\xi_1$ for all t, i.e. that $\Sigma_{t=1}^{\infty}a_t$ is convergent.

We conclude that if $\Sigma_{t=1}^{\infty}a_t$ is divergent, then Σ is efficient (since in this case there can be no improvement for Σ). The converse is also true; this will follow easily from our general results in the following sections.●

In the examples above, we saw that for a given reduced model Σ, if $(\xi_t)_{t\epsilon N}$ is an improvement, then $(-\xi_1,\xi_{t+1})\epsilon S_1 o\ldots o S_t$ for all t. In particular, we have that

$$\inf pr_1(S_1 o\ldots o S_t) \leq -\xi_1$$

for all t, where $pr_1 S=\{x\,|\,(x,y)\epsilon S\}$. Therefore, in the sequel we shall be interested in the sequence $(pr_1(S_1 o\ldots o S_t))_{t\epsilon N}$ of subsets of R.

These subsets can be given an interpretation in the spirit of the comments at the end of the previous section: If $x\epsilon pr_1(S_1 o\ldots o S_t)$, then $|x|$ units of the good can be had today against some repayment at date t+1 . Thus if $x\epsilon pr_1(S_1 o\ldots o S_t)$ for all t, $|x|$ units today can be traded in such a way that repayment will not be needed at any finite time t. Therefore Σ must be inefficient. On the other hand, if $x\notin pr_1(S_1 o\ldots S_\tau)$ for some τ, then $x\notin pr_1(S_1 o\ldots o S_t)$ for all $t\geq\tau$, and no amount y_t at date $t\geq\tau$ will compensate for $|x|$ at date 1. Consequently, Σ is efficient if for all $x<0$, $x\notin pr_1(S_1 o\ldots o S_\tau)$ holds for some τ.

The sets $S_1 o\ldots o S_t$ themselves will not do in this argument as shown by the following example:

EXAMPLE 3.8. Consider the sequence $(\tilde{S}_\nu)_{\nu\epsilon N}$, where

$$\tilde{S}_\nu = \{(x,y)\,|\,x+y\geq 0,\, 2^\nu x+y\geq 0,\, x\geq -1\}.$$

The sequence $(\tilde{S}_\nu)_{\nu\in N}$ converges to R^2_+ (in the closed convergence topology on the space of closed subsets of R^2, cf. Hildenbrand [1974]). However, the sequence $(pr_1\tilde{S}_\nu)_{\nu\in N}$ is constant with $pr_1\tilde{S}_\nu=[-1,\infty)$, all ν.

The reader may check that for the reduced model $\Sigma=(S_t)_{t\in N}$ with

$$S_t=\{(x,y)\,|\,x+y\geq 0,2x+y\geq 0,x\geq -t\},$$

all $t\in N$, one gets that $S_1 o\ldots oS_\nu=\tilde{S}_\nu$ (the argument follows that of Example 3.3). We conclude that Σ is inefficient even though $(S_1 o\ldots oS_t)\rightarrow R^2_+$.●

3.3. THE GENERAL EFFICIENCY CRITERION.

We return to the general case. As in the preceding section, for $S\subset R^2$ we let $pr_1S=\{x\,|\,(x,y)\in S\}$.

THEOREM 3.9. Let $\Sigma=(S_t)_{t\in N}$ be a reduced model where $S_t\in\Omega$, each $t\in N$. Then Σ is efficient if and only if

$$(3)\qquad \inf_{t\geq 1}\sup_{T\geq t}\inf pr_1(S_1 o\ldots oS_T)=0 .$$

Proof: Suppose that Σ is not efficient. Then there is an improvement $(\xi_t)_{t\in N}$ for Σ, i.e. $\xi_t\geq 0$ for all $t\in N$ and $\xi_t>0$ for some $t\in N$. Moreover, $(-\xi_t,\xi_{t+1})\in S_t$, all $t\in N$.

Let $r=\min\{t\,|\,\xi_t>0\}$. Then $-\xi_r\in pr_1(S_r o\ldots oS_T)$ for all $T\geq r$, consequently

$$0>-\xi_r \geq \sup_{T\geq r}\inf pr_1(S_r o\ldots oS_T)$$

contradicting (3).

Conversely, suppose that there is $r\geq 1$ and $a_r>0$ such that

$$\sup_{T\geq r}\inf pr_1(S_r o\ldots oS_T) = -a_r<0.$$

Choose $\bar{\xi}_r$ with $0<\bar{\xi}_r<a_r$. Then, by star-shapedness, the set

$A_T=\{y|(-\bar{\xi}_r,y)\in S_r o\ldots oS_T\}$

is non-empty for each $T\geq r$. Define a sequence $(\xi_t)_{t\in N}$ by $\xi_t=0$ for $t\leq r$, $\xi_r=\bar{\xi}_r$ and $\xi_t=\min A_T$ for $T>r$.

Clearly, $\xi_t\geq 0$ for all $t\in N$. Also, $\xi_t>0$ for $t\geq r$ since otherwise $(-\xi_r,0)\in S_r o\ldots oS_T$ for some $T\geq r$, a contradiction. We show that $(\xi_t)_{t\in N}$ is an improvement for Σ, i.e. that $(-\xi_t,\xi_{t+1})\in S_t$, all $t\in N$.

For $t\leq r-2$, $(-\xi_t,\xi_{t+1})=(0,0)\in S_t$ and $(-\xi_{r-1},\xi_r)\in R^2_+\subset S_{r-1}$. Moreover, $(-\xi_r,\xi_{r+1})\in S_r$ since $\xi_{r+1}=\min A_r$.

Suppose that we have shown $(-\xi_t,\xi_{t+1})\in S_t$ for $t\leq q$ where $q\geq r$. Then

$$(-\xi_r,\xi_{q+1})\in S_r o\ldots oS_q$$
$$(-\xi_r,\xi_{q+2})\in S_r o\ldots oS_q oS_{q+1}$$

(both by the definition of $(\xi_t)_{t\in N}$), and since $S_r o\ldots oS_{q+1}=$ $=(S_r o\ldots oS_q)oS_{q+1}$ there is $z\in R$ such that

$$(\xi_r,z)\in S_r o\ldots oS_q,\quad(-z,\xi_{q+2})\in S_{q+1}$$

and by the definition of ξ_{q+1}, $z\geq\xi_{q+1}>0$. Since S_{q+1} is star-shaped, we have

$$(\frac{\xi_{q+1}}{z})(-z,\xi_{q+2}) = (-\xi_{q+1},(\frac{\xi_{q+1}}{z})\xi_{q+2})\in S_{q+1}$$

and consequently,

$$(-\xi_r,(\frac{\xi_{q+1}}{z})\xi_{q+2})\in S_r o\ldots oS_{q+1}.$$

From the definition of ξ_{q+2} we conclude that $\xi_{q+1}=z$, whence $(-\xi_{q+1},\xi_{q+2})\in S_{q+1}$. By induction it follows that $(\xi_t,\xi_{t+1})\in S_t$ for all t, so Σ is not efficient.●

The criterion for efficiency given by Theorem 3.9 is, as can be seen, very general; it applies to all reduced models satisfying the rather weak regularity axioms discussed in Section 3.1. This generality is its strength but also its weakness, since the price which has to be paid is a high level of abstraction. For any particular reduced model it will demand additional computation to decide upon the question of efficiency.

Although the criterion for efficiency as presented here is new (see, however, Borglin and Keiding [1983]) a somewhat related construction, the so-called transfer function was introduced by Peleg and Yaari [1970]. However, the transfer function was introduced for another purpose than characterizing efficient programs, namely a description of different supports of production models (including models more general than those studied here) and the characterization of efficiency in terms of transfer functions is an immediate consequence of the definition of the latter concept, corresponding to a higher level of abstraction as compared to Theorem 3.9.

We shall not use the concept of transfer functions and the reader is referred to Peleg and Yaari [1970] for further details.

3.4. PARAMETRIC EFFICIENCY CRITERIA.

The result proved in the preceding section provides a characterization of efficient programs which applies to a very broad class of reduced models. However, due to its abstract character - composition of sets enter into the expression (3) - the theorem as it stands may be of limited use.

The examples treated in Section 3.2 give a hint of the direction in which to proceed in order to obtain more easily applied efficiency criteria. Namely, we shall consider reduced models $\Sigma=(S_t)_{t\in N}$ such that all S_t belong to a subsemigroup of Ω, the elements of which can be described by some parameters.

DEFINITION 3.10. Let Ω^o be a subsemigroup of Ω, and let $(M,*)$ be a semigroup. A <u>parametrization of Ω^o by M</u> is a semigroup homomorphism $\phi:M\rightarrow\Omega^o$ which is bijective. ●

EXAMPLE 3.11. The subsemigroup Ω^2 of Example 3.6 has a parametrization by (R_+,∇), where $\phi^{-1}(S)=b_S$ for $S\in\Omega^2$. The subsemigroup Ω^3 of Example 3.7 has a parametrization by $(\bar{R}_+,+)$, where we use the convention $\infty+a=a+\infty=\infty$ for all $a\in R$. The map $\phi:\bar{R}_+\rightarrow\Omega^3$ is given by $\phi(a)=S^a$.

It should be noted that the existence of a parametrization puts several restrictions on Ω^o; for example, if $(M,*)$ is commutative (i.e. $m_1*m_2=m_2*m_1$ for all $m_1,m_2\varepsilon M$), then so is Ω^o, since $\phi(a_1)o\phi(a_2)=$ $=\phi(a_1*a_2)=\phi(a_2*a_1)=\phi(a_2)o\phi(a_1)$.

To state our main theorem on parametric efficiency criteria, we need some further concepts and notation: Let R_Ω be the partial order relation on Ω defined by $SR_\Omega S'$ if $\inf pr_1 S\geq \inf pr_1 S'$. Similarly, let R_M be a partial order relation on M. Neither R_Ω nor R_M need to agree with the semigroup structure on Ω^o or M (although in our most prominent applications they do).

THEOREM 3.12. Let $\phi:M\longrightarrow\Omega^o$ be a parametrization such that for all $a,b\varepsilon M$, $aR_Mb \Leftrightarrow \phi(a)R_\Omega\phi(b)$. Further, let $\eta:M\longrightarrow R_+$ be a bijective map such that for all $a,b\varepsilon M$, $aR_Mb \Leftrightarrow \eta(a)\geq\eta(b)$.
Assume that $\Omega^o\subset\Omega$ is rich in the sense that $\{\inf pr_1 S|S\varepsilon\Omega^o\}$ is dense in some neighbourhood of 0 (in $[-\infty,0]$), and that $\eta(M)\subset R_+$ is unbounded.

Then a reduced model $\Sigma=(S_t)_{t\varepsilon N}$ with $S_t\varepsilon\Omega^o$ for all t is efficient if and only if

$$(4) \qquad \inf_{t\geq 1}\sup_{T\geq t}\phi(a_t*\dots*a_T) = \sup_{a\varepsilon M}\eta(a)$$

where $a_t = \phi^{-1}(S_t)$, $t\varepsilon N$.

Proof: Suppose that there is $t\varepsilon N$ and $a'>0$ such that

$$\inf pr_1(S_t o\dots oS_T) \leq -a'<0$$

for all $T\geq t$. By richness of Ω^o there are $S",S'\varepsilon\Omega^o$ such that

$$0 \geq \inf pr_1 S" > \inf pr_1 S' > \inf pr_1(S_t o\dots oS_T)$$

or $S"P_\Omega S'P_\Omega(S_t o\dots oS_T)$ for all T, where $SP_\Omega S'$, means that $SR_\Omega S'$, and not $S'R_\Omega S$. But then, with P_M defined in the same manner,

$$\phi^{-1}(S")P_M\phi^{-1}(S')P_M\phi^{-1}(S_t o\dots oS_T)=a_t*\dots*a_T$$

all $T\geq t$. Applying η and taking sup, we get

$$\sup_{a\varepsilon M}\eta(a) > \eta(\phi^{-1}(S')) > \sup_{T\geq t}\eta(a_t*\dots*a_T).$$

Therefore (4) implies that Σ is efficient.

Conversely, suppose that there is $t\epsilon N$ and $b\epsilon R$ such that

$$\sup_{T\geq t}\eta(a_t*\ldots*a_T) \leq b < \sup_{a\epsilon M}\eta(a).$$

Choose $a',a''\epsilon M$ such that $b<\eta(a')<\eta(a'')$. Then

$$a''P_M a' P_M (a_t*\ldots*a_T)$$

for all $T\geq t$, whence

$$\inf pr_1\phi(a'') > \inf pr_1\phi(a') > \inf pr_1(S_t o\ldots oS_T).$$

It follows that

$$\sup_{T\geq t}\inf pr_1(S_t o\ldots S_T) < 0,$$

so if Σ is efficient, then (4) holds.●

REMARK 3.13. Suppose that $\phi:M\longrightarrow\Omega^o$ is a surjective semigroup homomorphism (not necessarily injective), and that $\eta:M\longrightarrow R$ is such that $\phi(a)=\phi(b)$ implies $\eta(a)=\eta(b)$ for all $a,b\epsilon M$. Then $M/\sim_\phi$, where $a\sim_\phi b \Leftrightarrow \phi(a)=\phi(b)$, is a semigroup in the obvious way, and ϕ and η extend to maps $\phi:M/\sim_\phi\longrightarrow\Omega^o$ and $\eta:M/\sim_\phi\longrightarrow\Omega^o$. If ϕ,η and $M/\sim_\phi$ satisfy the assumptions of Theorem 3.12 we still obtain a parametric efficiency criterion.●

Below we consider some applications of Theorem 3.12. These applications will give an explanation of the somewhat complicated way, in which Theorem 3.12 was stated, in particular the presence of the map η.

EXAMPLE 3.6 (continued). Let $\phi:R_+\longrightarrow\Omega^o$ be defined by $\phi(b_S)=S$, let $R_M=\leq$ and let $\eta(a)=1/a$ for all $a\epsilon R_+$. Then Theorem 3.12 applies, and a reduced model $\Sigma=(S_t)_{t\epsilon N}$ with $S_t\epsilon\Omega^2$, all $t\epsilon N$, is efficient if and only if

$$\sup_{T\geq t}\frac{1}{\min\{b_t,\ldots,b_T\}} = \infty$$

for some $t\geq 1$, or, equivalently, if and only if

$$\liminf b_t=0.●$$

EXAMPLE 3.7 (continued). Define $\phi:R_+\to\Omega^3$ by $\phi(a)=S^a$, and let η be the identical map on R_+. Since $\inf pr_1 S^a=-1/a$ the richness assumption is fulfilled. Thus, by Theorem 3.12, a reduced model $S=(S_t)_{t\in N}$ is efficient if and only if

$$\inf_{t>1}\sup_{T\geq t}\Sigma_{\tau=t}^{T}a_\tau=\infty.\bullet$$

Thus, the efficiency criterion and its parametrized versions yield the standard types of criteria, the liminf-criterion and the divergent series criterion, as special cases. We note that the map η was inserted in the statement of Theorem 3.12 to allow for convergence to 0 as well as "convergence to infinity" as conditions for efficiency in particular cases.

Up to this point, we have treated only very special models. We shall see in the following chapter that these special cases actually provide for a parametrized characterization of efficiency in a much larger class.

3.5.* EFFICIENCY CRITERIA IN RELATED MODELS.

The treatment of efficiency/inefficiency in reduced models is the main topic of this book. However, we shall occassionally digress from this main topic in order to discuss some other problems which are rather closely related to it. Thus, we shall now give a short treatment of the production model with discounted utility maximization introduced in Section 2.5.

DEFINITION 3.14. Let $\Sigma=(S_t)_{t\in N}$ be a reduced model. A sequence $(\xi_t)_{t\in N}$ is a <u>σ-improvement</u> for Σ if

(i) $(-\xi_t,\xi_{t+1})\in S_t$

(ii) $\infty>\Sigma_{t=1}^{\infty}\xi_t>0$.

The reduced model $\Sigma=(S_t)_{t\in N}$ is <u>σ-efficient</u> if there is no σ-improvement for Σ. $\bullet$

By the Corollary to Theorem 2.4, a program $(x_t,y_t,c_t)_{t\in N}$ in the production model is optimal if and only if the corresponding reduced model is σ-efficient.

In order to characterize σ-efficient reduced models in a way comparable to that of Theorem 3.9 for efficient reduced models, we must introduce another composition rule on Ω.

LEMMA 3.15. Let * be the composition rule on Ω defined by

$$S*S' = \{(x,y)\,|\,\exists z_1,z_2\in R,(z_1,-z_2)\in S,(z_2,y)\in S',z_1+z_2=x\}$$

for $S,S'\in\Omega$. Then $(\Omega,*)$ is a semigroup.

<u>Proof</u>: First of all we show that * is indeed a composition rule on Ω, i.e. that $S*S'\in\Omega$.

$S*S'$ is closed: Let $(x^n,y^n)n\in N$ be a sequence with $(x^n,y^n)\in S*S'$, all n, converging to (x^o,y^o). Then there are z_1^n, z_2^n such that $(z_1^n,-z_2^n)\in S$, $(z_2^n,y^n)\in S'$, and $z_1^n+z_2^n=x^n$, all n. If both of the sequences $(z_1^n)_{n\in N}$ and $(z_2^n)_{n\in N}$ are bounded, it follows easily that $(x^o,y^o)\in S*S'$. If $z_1^n\to\infty$, then $z_2^n=x^n-z_1^n\to-\infty$, and from $(z_2^n,y)\in S'$ all n we get that $|z_2^n|^{-1}(z_2^n,y)\to(-1,0)\in S'$ as $n\to\infty$; a contradiction. If $z_1^n\to-\infty$ then $-z_2^n\to-\infty$ contradicting that $(z_1^n,-z_2^n)\in S$, so the sequence $(z_1^n)_{n\in N}$ is bounded and the arguments above show that also $(z_2^n)_{n\in N}$ is bounded.By star-shapedness $(|z_2^n|^{-1}z_2^n,|z_2^n|^{-1}y^n)\in S'$ for n sufficiently large. But then $(-1,0)\in S'$; a contradiction.

$S*S'$ satisfies (Ω1)-(Ω4): Clearly, $0\in S*S'$. Next, if $(x,y)\in S*S'$ and $u\in R_+^2$, then there are $z_1,z_2\in R$ with $(z_1,-z_2)\in S$, $(z_2,y)\in S'$ and $z_1+z_2=x$, and consequently $(z_1+u_1,-z_2)\in S$, $(z_2,y+u_2)\in S'$, from which we get that $(x+u_1,y+u_2)\in S*S'$.

Let $(x,y)\in S*S'$; if $y<0$, then for every $z_1,z_2\in R$ with $(z_1,-z_2)\in S$, $(z_2,y)\in S'$ we have $z_2>0$, thus $-z_2<0$ and $z_1>0$. It follows that $x=z_1+z_2>0$. Similarly, it can be shown that if $x<0$ then $y>0$. We conclude that $(S*S')\cap(-R_+^2)=\phi$. Finally, if $(x,y)\in S*S'$ and $\lambda\in[0,1]$, then it is easily seen that $(\lambda x,\lambda y)\in S*S'$. We conclude that * is indeed a composition on Ω.

Associativity of * follows since $(S*S')*S''$ equals

$\{(x,y)|\exists z_1,z_2,z_3\in R:(z_1,-z_2)\in S,(z_2,-z_3)\in S',(z_3,y)\in S'',z_1+z_2+z_3=x\}$

which again equals $S*(S'*S'')$.●

Now we can state the following characterization theorem for σ-efficiency, which is an almost direct translation of Theorem 3.9:

THEOREM 3.16. Let $\Sigma=(S_t)_{t\in N}$ be a reduced model where $S_t\in\Omega$, each $t\in N$. Then Σ is σ-efficient if and only if

(5) $$\sup_{T\geq 1}\inf pr_1(S_1*\ldots*S_T)=0 .$$

In the proof of Theorem 3.16 we shall need the following lemma:

LEMMA 3.17. Let $S_1,\ldots,S_T\in\Omega$ and suppose that $(x,y)\in S_1*\ldots*S_T$ is such that $x<0$. Then there are $z_1,\ldots,z_T\in -R_+$ such that $(z_t,-z_{t+1})\in S_t$, $t=1,\ldots,T-1$, $(z_T,y')\in S_T$, some $y'\in R$, and $\Sigma_{t=1}^T z_t=x$.

Proof: Let $z_1,\ldots,z_T\in R$ be such that $(z_t,-z_{t+1})\in S_t$, $t=1,\ldots,T-1$, $(z_T,y)\in S_T$ and $\Sigma_{t=1}^T z_t=x$. Suppose that $t>1$ and $z_t>0$. We conclude that

(a) $z_T\leq 0$, since otherwise $z_t>0$, all t, contradicting $\Sigma_{t=1}^T z_t=x<0$,

(b) either $z_t\leq 0$, all t, or there is τ with $1\leq\tau<T$ such that $z_t>0$ for $t\leq\tau$, $z_t\leq 0$ for $t>\tau$. Define $(z')_{t=1}^T$ by $z_t'=0$, $t\leq\tau$, $z_t'=z_t$, $t>\tau$. Then

$$(z_t',-z_{t+1}')\in S_t$$

for $t<\tau$, since $0\in S$, all $S\in\Omega$, $(z_t',-z_{t+1}')\in R_+^2\subset S_t$, and

$$(z_t',-z_{t+1}')=(z_t,-z_{t+1})\in S_t$$

for $t>\tau$. Finally, $\Sigma_{t=1}^T z_t'\leq\Sigma_{t=1}^T z_t=x$. Let $\lambda=(\Sigma_{t=1}^T z_t')^{-1}(\Sigma_{t=1}^T z_t)$ and define $(z'')_{t=1}^T$ by $z_t''=\lambda z_t'$. Then $(z'')_{t=1}^T$ has the properties stated in the lemma.●

Proof of Theorem 3.16: The "only if" part is straightforward and is left to the reader.

"If": Suppose that there is $a>0$ such that

$$\inf pr_1(S_1*\ldots*S_T) < -a$$

for all T. By Lemma 3.17, for each T there are $z_1^T,\dots z_T^T \le 0$ such that $\Sigma_{t=1}^T z_t^T = -a$, $(z_t^T, -z_{t+1}^T) \in S_t$, $t = 1,\dots,T-1$, and $(z_T^T, y^T) \in S_T$ for some y^T. Clearly, $|z_t| \le a$ for all t and T. Define for each T the sequence $(\xi_t^T)_{t \in N}$ by $\xi_t^T = -z_t^T$, $t \le T$, $\xi_t = 0$, $t > T$. Using Cantor's diagonal process we can find a subsequence $(T_\nu)_{\nu \in N}$ of $(1,2,\dots,T,\dots)$ such that for each $t \in N$, $(\xi_t^{T_\nu})_{\nu \in N}$ converges to some ξ_t^o. Since $(-\xi_t^{T_\nu}, \xi_{t+1}^{T_\nu}) \in S_t$ for all ν sufficiently large, we have that $(-\xi_t^o, \xi_{t+1}^o) \in S_t$, all t. Finally, since $\Sigma_{t=1}^\infty \xi_t^{T_\nu} = a$ for all ν, we get by Fatou's lemma that $\Sigma_{t=1}^T \xi_t^o = a$. Consequently, $(\xi_t^o)_{t \in N}$ is a σ-improvement.●

Theorem 3.16 has a corollary which will be useful in our further treatment of this model:

COROLLARY. Let Σ be a reduced model, $(\xi_t)_{t \in N}$ a sequence such that $(-\xi_t, \xi_{t+1}) \in S_t$, all t, and $\Sigma_{\tau=1}^T \xi_\tau > 0$ for all T. Then Σ is not σ-efficient.

Proof: Clearly, $\inf_{T \ge 1} pr_1(S_1 * \dots * S_T) \le -\xi_1 < 0$ for all T.●

CHAPTER 4: MEASURES OF CURVATURE AND GENERAL EFFICIENCY CRITERIA

In the previous chapter we obtained a complete characterization of efficient models $\Sigma=(S_t)_{t\in N}$ where each S_t belonged to a small class of well-behaved sets.

In this chapter we give two extensions of this result, thereby significantly extending the applicability of the characterization. First of all, we consider reduced models with arbitrary supports, and secondly, we use the reduced models considered hitherto for approximation of arbitrary reduced models.

Having thus pursued the derivation of parametric efficiency criteria as far as possible in the context of reduced models, we return to the original production and consumption models treated in Chapter 1 and 2 in order to interpret our results and compare them with those of the literature. Finally, we give a short treatment of some other models in which our approach can yield some results.

4.1. REDUCED MODELS WITH ARBITRARY SUPPORT.

In the applications of the result on parametric efficiency criteria (Theorem 3.12) we gave some examples of a rather simple character;

specifically, the reduced models $\Sigma=(S_t)_{t\epsilon N}$ in Examples 3.6 and 3.7 have support $(1,1,...)$. Now we want to consider reduced models with arbitrary support $(p_t)_{t\epsilon N}$.

Define the family $\Omega^4 \subset \Omega$ of sets

$$S(a;p_1,p_2)=\{(x,y)\,|\,p_1x+p_2y\geq 0,\ p_1x+p_2y+a(p_1x)(p_2y)\geq 0\}$$

where $a\epsilon R_+$, $p_1,p_2 \epsilon \dot{R}_+$, together with the set R^2_+.

Note that $S(a;p_1,p_2)=S(a/\lambda;\lambda p_1,\lambda p_2)$ for every $\lambda>0$, thus the same set may arise for different triples $(a;p_1,p_2),(a';p_1',p_2')$. If we introduce the convention $S(\infty;p_1,p_2)=R^2_+$ for all $p_1,p_2\epsilon\dot{R}_+$, we can write Ω^4 as $\Omega^4=\{S(a;p_1,p_2)\,|\,(a,p_1,p_2)\epsilon\bar{R}_+x\dot{R}_+x\dot{R}_+\}$.

The reader might check at this point that

$$S(a;p_1,p_2)oS(a';p_1',p_2')=S(a+\frac{p_1'}{p_2}\,a';p_1,\frac{p_2}{p_1'}\,p_2')$$

for all $(a,p_1,p_2),(a',p_1',p_2')\epsilon\bar{R}_+x\dot{R}_+x\dot{R}_+$, so that Ω^4 is a subsemigroup of Ω. Now the map $(a,p_1,p_2)\longrightarrow S(a;p_1,p_2)$ is not a parametrization since it fails to be injective. However, in view of Remark 3.13, this is of minor importance, and since it can be checked that the remaining conditions of Theorem 3.12 are fulfilled, we obtain a parametric efficiency criterion for reduced models $\Sigma=(S_t)_{t\epsilon N}$ with $S_t\epsilon\Omega^4$, $t\epsilon N$.

To avoid calculations, we follow another approach exploiting the results obtained for Ω^3 (Example 3.7).

We need some general concepts:

DEFINITION 4.1. Let Ψ be the set of reduced models.

(i) A <u>morphism in Ψ</u> is a family $G=(g_t)_{t\epsilon N}$ of maps $g_t:R\longrightarrow R$ such that each g_t is an order-preserving linear homomorphism.

(ii) For G a morphism and $\Sigma\epsilon\Psi$ a reduced model, G takes Σ to

$$G(\Sigma)=(S_t')_{t\epsilon N}$$

defined by

$$S'_t=\{(x',y')\mid\exists(x,y)\in S_t, x'=g_t(x), y'=g_{t+1}(y)\}. \bullet$$

REMARK 4.2.

(1) It is easily seen that $G(\Sigma)$ is a reduced model.

(2) A homomorphism $g:R\to R$ can be identified with a number $g\in R$ such that $g(x)=gx$ for all $x\in R$. If g is order preserving, then $g>0$. Our use of the rather complicated terminology is motivated by the use of morphisms in Part II.

(3) If $G=(g_t)_{t\in N}$ is a morphism, then each g_t is invertible, so $G^{-1}=(g_t^{-1})_{t\in N}$ is a morphism, and $GoG^{-1}=G^{-1}oG=id_\Psi$, where id_Ψ is the identical morphism with $g_t=id_R$, each t.

LEMMA 4.3. Let Σ be a reduced model, G a morphism. Then Σ is efficient if and only $G(\Sigma)$ is efficient.

Proof: Suppose that Σ is not efficient and let $(\xi_t)_{t\in N}$ be an improvement. Then $(g_t(\xi_t))_{t\in N}$ is an improvement for $G(\Sigma)$, which consequently is inefficient. The conclusion of the lemma now follows from Remark 4.2(3). $\bullet$

Now, let $\Sigma=(S_t)_{t\in N}$ be a reduced model with $S_t=S(a_t;p_1^t,p_2^t)\in\Omega^4$ for all $t\in N$. Suppose further that Σ is weakly efficient and has support $(p_t)_{t\in N}$. We assume that the parameters a_t,p_1^t,p_2^t have been chosen such that $(p_1^t,p_2^t)=(p_t,p_{t+1})$ for each $t\in N$.

Define the morphism $G=(g_t)_{t\in N}$ by $g_t(x)=(p_t)^{-1}x$ for $x\in R$. Then $G(\Sigma)=(S^{a_t})_{t\in N}$ with $S^{a_t}\in\Omega^3$, each $t\in N$. Applying Example 3.7 and Lemma 4.3, we have:

THEOREM 4.4. Let $\Sigma=(S_t)_{t\in N}$ be a reduced model with support $(p_t)_{t\in N}$ such that $S_t=S(a_t;p_t,p_{t+1})\in\Omega^4$, each $t\in N$. Then Σ is efficient if and only if

$$\inf_{t\geq 1}\sup_{T\geq t}\Sigma_{\tau=t}^{T}a_\tau = \infty. \bullet$$

Thus, the efficiency criterion in "divergent series" form extends to reduced models with arbitrary supports.

By the same methods, the efficiency criterion of Example 3.6 may be extended to reduced models with arbitrary supports. We omit the details and state the final result:

THEOREM 4.5. Let $\Sigma=(S_t)_{t\epsilon N}$ be a reduced model with support $(p_t)_{t\epsilon N}$ and suppose that

$$S_t=\{(x,y)|p_t x+p_{t+1}y\geq 0, x\geq -b_t\}$$

for each $t\epsilon N$. Then Σ is efficient if and only if

$$\liminf_{t\to\infty} p_t b_t=0. \bullet$$

4.2. EFFICIENCY CRITERIA BY APPROXIMATION AND MEASURES OF CURVATURE.

In this section we give another extension of the results on parametric efficiency criteria. The key to the following results is the next lemma:

LEMMA 4.6. Let Σ and Σ' be reduced models and suppose that $\Sigma\subset\Sigma'$ in the sense that $S_t\subset S'_t$, each $t\epsilon N$. Then Σ is efficient if Σ' is efficient.

Proof: Suppose that there was an improvement $(\xi_t)_{t\epsilon N}$ for Σ. Then, since $S_t\subset S'_t$, each $t\epsilon N$, we have that $(\xi_t)_{t\epsilon N}$ is also an improvement for Σ', a contradiction.$\bullet$

The idea of efficiency criteria by approximation is the following: If $\Sigma=(S_t)_{t\epsilon N}$ is a reduced model with support $(p_t)_{t\epsilon N}$, and $\Sigma\subset\Sigma'$, where $S_t= =S(a';p_t,p_{t+1})$, all $t\epsilon N$, and $\Sigma_{t=1}^{\infty}a'_t=\infty$, then Σ', and therefore Σ, is efficient. Conversely, if Σ is efficient and $\Sigma''\subset\Sigma$, where $\Sigma''= =S(a'';p_t,p_{t+1})$, then Σ'' is efficient, consequently $\Sigma_{t=T}^{\infty}a''_t=\infty$.

In the above argument, the given reduced model Σ was approximated from within (by Σ'') and from the outside (by Σ') in a "pointwise" fashion, i.e. we have

(1) $S''_t\subset S_t\subset S'_t$

for each $t\epsilon N$. Clearly, the sharpest result is obtained when Σ''_t (Σ'_t) is chosen as large (small) as possible in (1).

Note that the concept of a set S''_t or S'_t chosen as close as possible (in the above sense) to S_t is intimately related to the usual definition of the curvature of bdS_t at 0, whereby the ball is found whose boundary is the closest possible to that of S_t at the point 0. This provides a motivation for the following terminology:

DEFINITION 4.7. Let $S\epsilon\Omega$ and $p=(p_1,p_2)\epsilon\dot{R}_+x\dot{R}_+$ a support for S. The <u>outer curvature</u>, $m_p(S)$, of S is the extended real number

$$m_p(S)=\sup\{a|S\subset S(a;p_1,p_2)\}$$

and the <u>inner curvature</u>, $M_p(S)$, of S is the extended real number

$$M_p(S)=\inf\{a|S(a;p_1,p_2)\subset S\}. \bullet$$

REMARK 4.8. It will be useful in the following to note already at this point that

$$\cap_{a<m_p(S)}S(a;p_1,p_2)=S(m_p(S);p_1,p_2).$$

Indeed, if $S\subset S(a;p_1,p_2)$ for all $a<m_p(S)$, then each $(x,y)\epsilon S$ satisfies

$$p_1x+p_2y+a(p_1x)(p_2y) \geq 0$$

for all $a<m_p(S)$, and therefore also

$$p_1x+p_2y+m_p(S)(p_1x)p_2y) \geq 0.$$

Thus $S\subset S(m_p(S);p_1,p_2)$ and the conclusion follows. A similar argument will show that $S(M_p(S);p_1,p_2)\subset S$, so that

$$S(M_p(S);p_1,p_2)=cl\cup_{a>M_p(S)}S(a;p_1,p_2).$$

Now we can state the following result:

THEOREM 4.9. Let $\Sigma=(S_t)_{t\epsilon N}$ be a reduced model with support $(p_t)_{t\epsilon N}$. Write $p^t=(p_t,p_{t+1})$, $m_t=m_{p^t}(S_t)$ and $M_t=M_{p^t}(S_t)$ for each t. Then:

(a) If $\Sigma_{t=T}^{\infty} m_t$ is divergent, all T, then Σ is efficient.

(b) If $\Sigma_{t=T}^{\infty} M_t$ is convergent, some T, then Σ is inefficient.

__Proof__: The theorem follows by a straightforward application of Theorem 4.4 and 4.6 together with Remark 4.8. The details are left to the reader.●

Theorem 4.9 gives a criterion for efficiency expressed in terms of (inner and outer) curvature as well as supporting prices. Clearly, we have

$$m_p(S) \leq M_p(S)$$

for every $S \in \Omega$; the inequality may very well be proper. Thus, if S has several supports at 0 (i.e. S has non-smooth boundary), then $M_p(S)=\infty$, while $m_p(S)$ may be finite.

It follows that Theorem 4.9 does not provide a complete characterization of efficiency, since there are reduced models $\Sigma=(S_t)_{t \in N}$ with $\Sigma_{t=1}^{\infty} m_t$ convergent and $\Sigma_{t=1}^{\infty} M_t$ divergent.

EXAMPLE 4.10. Let $\Sigma=(S_t)_{t \in N}$ be the reduced model given by

$$S_t=\{(x,y) \mid x+y \geq 0,\ x \geq -\frac{1}{t}\}$$

for each t. It is easily checked that $M_t=1/t$ while $m_t=0$, each t. Therefore, $\Sigma_{t=1}^{\infty} M_t$ is divergent, $\Sigma_{t=1}^{\infty} m_t$ convergent. It is easily seen, using the results in Example 3.6, that Σ is efficient.

Let $\Sigma'=(S'_t)_{t \in N}$ be the reduced model given by, for each t,

$$S'_t=\{(x,y) \mid x+y \geq 0,\ x \geq -1\ ,\ x+y+2^{-t}xy \geq 0\}$$

Here $M_t=1$ and $m_t=2^{-t}$, so again $\Sigma_{t=1}^{\infty} M_t$ is divergent, $\Sigma_{t=1}^{\infty} m_t$ convergent, but this time Σ' is inefficient (by the Theorem 4.14 to follow).●

Theorem 4.9 has the following

COROLLARY. In the notation of Theorem 4.9 , suppose that M_t is finite, all t, and that there is K>0 such that

$$m_t \leq M_t \leq K m_t$$

for all t. Then Σ is efficient if and only if $\Sigma_{t=1}^{\infty} m_t$ is divergent.

Proof: If $\Sigma_{t=1}^{\infty} m_t$ is divergent, then $\Sigma_{t=T}^{\infty} m_t$ is divergent, all T, and Σ is efficient. If $\Sigma_{t=1}^{\infty} m_t$ is convergent, then so is $\Sigma_{t=1}^{\infty} M_t$, and Σ is not efficient.●

4.3.* LOCAL APPROXIMATIONS.

For the applications of Theorem 4.9 it turns out that a slight sharpening of the results is useful. It should be noted that the inner and outer approximations of a set $S\epsilon\Omega$ defined above can be considered as global approximations, and this is actually more than is needed, since everything interesting takes place in the region $\{(x,y)|x\leq 0\}$. Therefore we need only approximate S in this region.

We prefer to state this concept of local approximation in full generality:

DEFINITION 4.11. Let $S\epsilon\Omega$, $p=(p_1,p_2)\epsilon\dot{R}_+ x\dot{R}_+$ a support for S and $B\subset R^2$ an arbitrary set. The <u>outer curvature of S relative to B</u> is the extended real number

$$m_{p,B}(S)=\sup\{a|[S\cap B]\subset[S(a;p_1,p_2)\cap B]\}$$

and the <u>inner curvature of S relative to B</u> is the extended real number

$$M_{p,B}(S)=\inf\{a|[S(a;p_1,p_2)\cap B]\subset[S\cap B]\}. \bullet$$

THEOREM 4.12. Let $\Sigma=(S_t)_{t\epsilon N}$ be a reduced model with support $(p_t)_{t\epsilon N}$, and let $(B_t)_{t\epsilon N}$ be a sequence of subsets of R^2. If $\Sigma_{t=T}^{\infty} m_{p^t,B_t}(S_t)$ is divergent, all T, then there is no improvement $(\xi_t)_{t\epsilon N}$ for Σ with $(-\xi_t,\xi_{t+1})\epsilon B_t$, all t.

Proof: Suppose that $(\xi_t)_{t\epsilon N}$ is an improvement for Σ with $(-\xi_t,\xi_{t+1})\epsilon B_t$. Choose a number $\delta>0$ and let a_t be such that, for $t\epsilon N$

$$a_t\geq m_{p^t B_t}(S_t)-\delta \text{ and}$$

$$S\cap B_t\subset S(a_t;p_t,p_{t+1})$$

From the first inequality we get that $\Sigma_{t=T}^{\infty} a_t$ is divergent, all T, and from the second expression we have that $(\xi_t)_{t\epsilon N}$ is an improvement for $\Sigma'=(S(a_t;p_t,p_{t+1}))_{t\epsilon N}$. Thus, we have a contradiction.●

We shall state a converse of Theorem 4.12 in a somewhat less general form which will be useful for later applications. First of all we need a lemma:

LEMMA 4.13. Let $\Sigma=(S_t)_{t\epsilon N}$ be a reduced model with support $(p_t)_{t\epsilon N}$, and suppose that $S_t=S(a_t;p_t,p_{t+1})\epsilon\Omega^4$ for each t. If Σ is inefficient, then for each $\lambda>0$ there is an improvement $(\xi_t)_{t\epsilon N}$ such that the sequence $(p_t\xi_t)_{t\epsilon N}$ converges to λ.

<u>Proof</u>: By Theorem 4.9 there is $T\geq 1$ such that $\Sigma_{t=T}^{\infty} a_t$ is convergent, thus $a_t \longrightarrow 0$ for $t\longrightarrow\infty$. Let $\lambda>0$ be arbitrary and choose $T'\geq T$ such that $\lambda^{-1}-\Sigma_{t=T'}^{\infty} a_t>0$.

Define the sequence $(\xi_t)_{t\epsilon N}$ by $\xi_t=0$, $t<T'$, $\xi_{T'}=(\lambda^{-1}-\Sigma_{t=T'}^{\infty} a_t)^{-1}$ and

(2) $$\xi_t=(\Sigma_{\tau=T'}^{t-1} a_\tau+\xi_{T'})^{-1}$$

for $t>T'$. We show that $(\xi_t)_{t\epsilon N}$ satisfies the conclusion of the lemma:

For $t\longrightarrow\infty$ we have that

$$\xi_t\longrightarrow(\Sigma_{\tau=T'}^{\infty}\frac{a_\tau}{\xi_{T'}})^{-1} = (\Sigma_{\tau=T'}^{\infty} a_\tau+\lambda^{-1} - \Sigma_{\tau=T'}^{\infty} a_\tau)^{-1} = \lambda$$

so the sequence $(\xi_t)_{t\epsilon N}$ converges to λ. To show that $(\xi_t)_{t\epsilon N}$ is an improvement, it suffices to show that $(-\xi_{T'},\xi_t)\epsilon S_{T'}o\ldots oS_{t-1}$ for all $t>T'$. But this follows easily from (2) since we have

$$-\frac{1}{\xi_{T'}}+\frac{1}{\xi_t} = \Sigma_{\tau=T'}^{t-1} a_\tau$$

or

$$\xi_{T'}+\xi_t-(\Sigma_{\tau=T'}^{t-1} a_\tau)\xi_{T'}\xi_t=0$$

In the case of arbitrary support $(p_t)_{t\epsilon N}$, let $G=(g_t)_{t\epsilon N}$ be the morphism with $g_t(x)=p_t x$, each $x\epsilon R$. Then $G(\Sigma)$ has support $(1,1,\ldots)$ and G^{-1} takes an improvement $(\xi_t)_{t\epsilon N}$ in $G(\Sigma)$ with $\xi_t\longrightarrow\lambda$ to an improvement $(\xi_t)_{t\epsilon N}$ in Σ with $p_t\xi_t\longrightarrow\lambda$.●

Now we have the follwing converse of Theorem 4.12:

THEOREM 4.14. Let $\Sigma=(S_t)_{t\epsilon N}$ be a reduced model with support $(p_t)_{t\epsilon N}$, let $b_t=-\inf\{x\mid(x,y)\epsilon S_t\}$, each t, and let $a=\inf p_t b_t>0$. Finally, define sets

$$D_t(\lambda,\lambda')=\{(x,y)\mid\lambda a\le -p_t x\le\lambda' a\}$$

for $0\le\lambda<\lambda'$. If for some λ,λ' with $0\le\lambda<\lambda'$ we have

$$\Sigma_{t=1}^{\infty} M_{p^t,D_t(\lambda,\lambda')}(S_t) < \infty$$

then S is inefficient.

<u>Proof</u>: The reduced model $\Sigma'=(S_t')_{t\epsilon N}$ with

$$S_t'=S(a';p_t,p_{t+1})$$
$$a_t'=M_{p^t,D_t(\lambda,\lambda')}(S_t)$$

for $t\epsilon N$, is inefficient. By Lemma 4.13, there is an improvement $(\xi_t)_{t\epsilon N}$ for Σ' such that $p_t\xi_t\to\lambda' a$. Since clearly $p_t\xi_t\le p_{t+1}\xi_{t+1}$ for all t, there is T such that $p_t\xi_t\ge\lambda a$ for all $t\ge T$.

Let $(\bar{\xi}_t)_{t\epsilon N}$ be the improvement for Σ' defined by $\bar{\xi}_t=0$, $t<T$, $\bar{\xi}_t=\xi_t$, $t\ge T$. Then $(\bar{\xi}_t)_{t\epsilon N}$ is also an improvement for Σ, which therefore is inefficient.●

4.4. MEASURES OF CURVATURE IN THE PRODUCTION MODEL.

Having obtained in the previous sections efficiency criteria for reduced models in a rather general form, we now turn to a discussion of these results in the context of the original models.

Clearly a central role has been played by the measure of curvature introduced in Section 4.3. Following our discussion in Chapter 2, we note that finding inner (outer) curvature of a reduced model derived from a program in a production model amounts to approximation of f_t at x_t by a hyperbola from below (above), as shown in Figure 4.1.

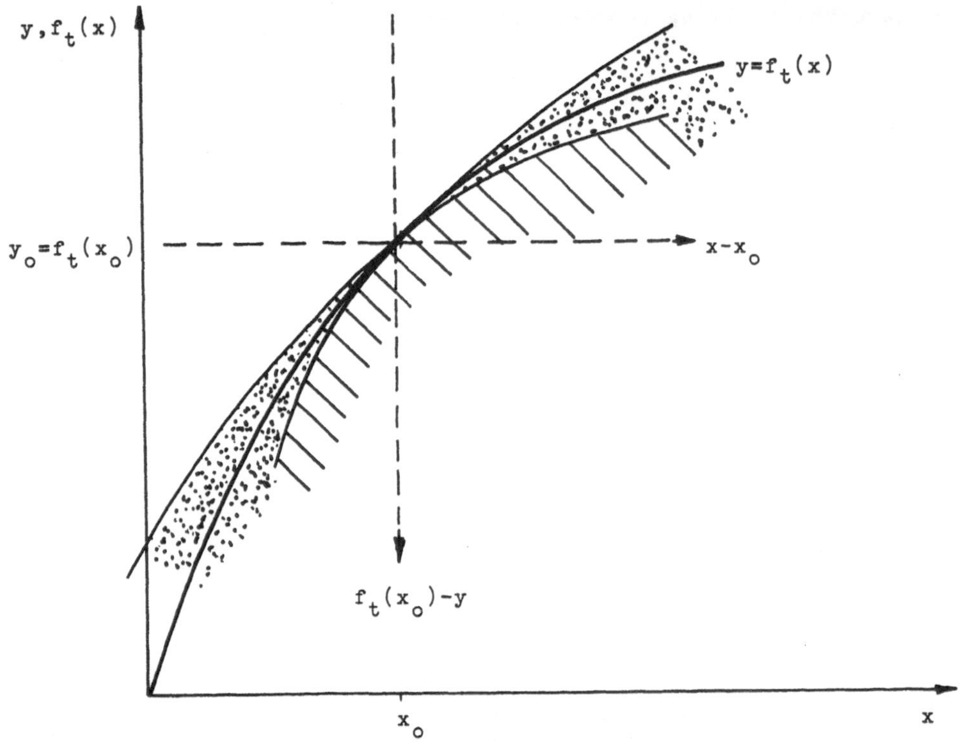

FIGURE 4.1

This way of measuring curvature may not be particularly convincing from a geometric or a function-theoretical point of view. However, this need not concern us too much. We have argued in the preceding chapters that this particular measure of curvature has its principal advantage in its tractability. Moreover, as will be shown shortly, other measures can be "translated" rather easily to this one.

4.4.1. Parabolic Measures of Curvature.

An alternative suggested by the figure is to use parabola instead of hyperbola for the approximation of f_t at x_t. This will have the advantage of yielding a rotation-invariant measure of curvature, whereas the measure considered hitherto depends on the support in a non-trivial way.

To avoid excessive notation, we treat this parabolic measure of curvature in the context of the reduced model, even though the intuitive argument for it as well as its actual use in the literature is connected with the original production model.

For $z=(x,y)\epsilon R^2$ and $p=(p_1,p_2)\epsilon \dot{R}_+ x\dot{R}_+$, let z^p denote the projection of z on p and let $z^{\perp}=z-z^p$. Define the family $S_p^{\perp}(\cdot)$ of convex sets by

$$S_p^{\perp}(b)=\{z\epsilon R^2 \mid \|z^p\|\geq b\|z^{\perp}\|^2\}.$$

Note that $S_p^{\perp}(b)$ does not belong to Ω since it fails to satisfy Ω2. This, however, is of minor importance since anyway the family $S_p^{\perp}(\cdot)$ does not admit any simple semigroup structure.

For $S\epsilon\Omega$ and p a support for S, let

$$m_p^{\perp}(S)=\sup\{b \mid S\subset S_p^{\perp}(b)\}$$

$$M_p^{\perp}(S)=\inf\{b \mid S_p^{\perp}(b)\subset S\}$$

(both being extended real numbers) be <u>outer and inner parabolic curvature</u>.

The following lemma gives the relation between m and $m^{\perp}$:

LEMMA 4.15. $S_p^{\perp}(b)\subset S(\frac{2b}{\|p\|};p_1,p_2)$.

<u>Proof</u>: For each $z=(x,y)\epsilon S_p^{\perp}(b)$, we have

$$\|z^p\|=\frac{1}{\|p\|}(p_1x+p_2y)$$

and

$$\|z^{\perp}\|^2=\|z\|^2-\|z^p\|^2=$$

$$=x^2+y^2-\frac{1}{\|p\|^2}((p_1x)^2+(p_2y)^2+2(p_1x)(p_2y))=$$

$$=x^2(1-\frac{p_1^2}{\|p\|^2})+y^2(1-\frac{p_2^2}{\|p\|^2})-\frac{1}{\|p\|^2}(p_1x)(p_2y)=$$

$$=\frac{1}{\|p\|^2}(p_1y-p_2x)^2$$

so from $\|z^p\|\geq b\|z^{\perp}\|^2$ we get

(3) $$p_1x+p_2y \geq \frac{b}{\|p\|}(p_1y-p_2x)^2 \geq \frac{-2b}{\|p\|}(p_1x)(p_2y),$$

i.e., $z \in S(\frac{2b}{\|p\|};p_1,p_2)$.●

As a corollary of Lemma 4.15 we get that $m_p(S) \geq 2m_p^{\perp}(S)/\|p\|$. A converse of the lemma (asserting that $S(a;p_1,p_2)$ is contained in $S_p^{\perp}(b)$ for some $b>0$) cannot be obtained since the sets $S_p^{\perp}(b)$ do not satisfy the monotonicity assumption Ω2. However, we have the following:

LEMMA 4.16. Let $z=(x,y)$ be such that $|p_2y|>|p_1x|$ and

$$p_1x+p_2y \geq \frac{b}{\|p\|}(\frac{p_2}{p_1} + \frac{p_1}{p_2})^2(p_2y)^2.$$

Then $z \in S_p^{\perp}(b)$.

Proof: From

$$(p_2y)^2 \geq (-\frac{1}{2}p_1x + \frac{1}{2}p_2y)^2$$

and

$$(\frac{p_2}{p_1} + \frac{p_1}{p_2})(-p_1x+p_2y) \geq (p_1y-p_2x)$$

we get that

$$p_1x+p_2y \geq \frac{b}{\|p\|}(p_1y-p_2x)^2$$

and thus, by (3), $z \in S_p^{\perp}(b)$.●

Using Lemma 4.15 and 4.16, we get the following reformulation of Theorem 4.9:

THEOREM 4.17. Let $\Sigma=(S_t)_{t \in N}$ be a reduced model with support $(p_t)_{t \in N}$. Let $p^t=(p_t,p_{t+1})$ and $m_t^{\perp}=m_{p^t}^{\perp}(S_t)$, $M_t^{\perp}=M_{p^t}^{\perp}(S_t)$, $t \in N$.

(a) If for all T, $\Sigma_{t=T}^{\infty}\frac{m_t^{\perp}}{\|p^t\|}$ is divergent, then Σ is efficent.

(b) If for some T, $\Sigma_{t=T}^{\infty}\frac{M_t^{\perp}}{\|p^t\|}(\frac{p_{t+1}}{p_t} + \frac{p_t}{p_{t+1}})^2$ is convergent,

then Σ is inefficient.●

Theorem 4.17 shows that the results obtained using the hyperbolic measure of curvature carry over to results with the present types of measures. The precise form of the efficiency criteria may be less important. It should be noted, however, that for models where the parabolic curvatures are uniformly bounded ($k \leq m_t^{\perp} \leq M_t^{\perp} \leq K$) and prices do not fluctuate too much ($c \leq p_t / p_{t+1} \leq C$, $c \leq p_{t+1}/p_t \leq C$, for some constants c and C), we get that efficiency depends on whether or not the series

$$\Sigma_{t=1}^{\infty} \frac{1}{p_t}$$

is divergent. This is the Cass criterion referred to in Section 1.1.

4.4.2. Benveniste's Measure of Curvature.

Benveniste [1976b] defines certain measures of curvature of convex sets in R^{2n}. We give a short account of these measures in the context of reduced models. It will, however be convenient to assume that the sets in question do not contain 0.

For $S \varepsilon \Omega$ with support $p=(p_1,p_2)$ and u a vector in R^2 with $u_1>0$, let $S'=S+\{u\}$ and define the <u>coefficient of strictness</u> of S' at u as $\sup\lambda$, where λ is such that for all $(x,y) \varepsilon S'$,

$$p_1(x-u_1)+p_2(y-u_2) \geq \frac{\lambda}{p_1 u_1} (p_1(x-u_1))^2$$

Note that, contrary to the situation with hyperbolic measures, this measure of curvature is independent of scalar multiplication of prices.

The coefficient of strictness is in our terminology an outer curvature. The counterpart of inner curvature is the <u>coefficient of smoothness</u>, defined as the infimum over all λ such that

$$p_1(x-u_1)+p_2(y-u_2) \geq \frac{\lambda}{p_1 u_1}(p_1(x-u_1))^2 \text{ implies } (x,y) \varepsilon S'$$

The coefficients of strictness and smoothness are closely related to the parabolic curvatures considered in Section 4.4.1. However, they are not identical with those. We leave a detailed comparison to the reader, since anyway, a measure of curvature which is very similar will be treated in some detail in the next section.

4.4.3. Mitra's Condition S.

We conclude this short review of alternative measures of curvature and corresponding efficiency criteria, having appeared in the literature on the production model, with a discussion of a production model treated in Mitra [1979b].

Let $(f_t)_{t\epsilon N}$ be a production model with $f_t=f$, all $t\epsilon N$, where f is strictly increasing and differentiable, and suppose that all feasible programs $(x_t,y_t,c_t)_{t\epsilon N}$ satisfy the following

CONDITION S. For some $0<m\leq M<\infty$, $N<\infty$, and $0<\lambda\leq 1$, there exists a function $\mu(x)$ for $x\geq 0$, such that

(a) $0 \leq \mu(x_t) \leq N$ for $x_t>0$, $t\geq 0$, and

(b) $m\xi \dfrac{\mu(x_t)}{x_t} \leq \dfrac{(f(x_t)-f(x_t-\xi))}{\xi f'(x_t)} -1 \leq M\xi \dfrac{\mu(x_t)}{x_t}$

for $0<\xi<\lambda x_t$, $t\geq 0$.

Mitra shows that a feasible program $(x_t,y_t,c_t)_{t\epsilon N}$ satisfying Condition S is inefficient if and only if

$$\inf p_t x_t > 0 \text{ and } \Sigma_{t=1}^{\infty} \frac{\mu(x_t)}{p_t x_t} < \infty.$$

We now show how this result can be obtained using our approach. First of all, if the program $(x_t,y_t,c_t)_{t\epsilon N}$ satisfies Condition S, then

$$(4) \qquad 1 \leq \frac{(p_{t+1}(f(x_t)-f(x_t-\xi))}{p_t\xi} \leq 1+\lambda MN$$

for $0<\xi<x_t$. Indeed, inserting $f'(x_t)=p_t/p_{t+1}$ in the right-hand inequality in (b) yields

$$\frac{p_{t+1}(f(x_t)-f(x_t-\xi))}{p_t\xi} -1 \leq M\xi\frac{\mu(x_t)}{x_t} \leq \lambda M\mu(x_t),$$

and using (a) we get the right-hand inequality of (4). Similar reasoning shows that the left-hand inequality of (4) follows from the left-hand inequality of (b).

Next, rewrite the left- and right-hand inequalities in (b) as

$$(5) \qquad p_{t+1}(f(x_t)-f(x_t-\xi))-p_t\xi \geq m\,\frac{\mu(x_t)}{p_t x_t}(p_t\xi)^2$$

$$(6) \qquad p_{t+1}(f(x_t)-f(x_t-\xi))-p_t\xi \leq M\,\frac{\mu(x_t)}{p_t x_t}(p_t\xi)^2$$

for $0<\xi<\lambda x_t$. Expressions (4) and (5) together give

$$p_{t+1}(f(x_t)-f(x_t-\xi))-p_t\xi \geq \frac{m}{1+\lambda MN}\,\frac{\mu(x_t)}{p_t x_t}\,p_t\xi(f(x_t)-f(x_t-\xi))$$

from which it follows that

$$m_{p^t,B_t}(S_t) \geq \frac{m}{1+\lambda MN}\,\frac{\mu(x_t)}{p_t x_t}$$

where S_t is defined as in Chapter 2 and $B_t=\{(x,y)\,|\,0>x>-\lambda x_t\}$.

Furthermore, by (6),

$$M\,\frac{\mu(x_t)}{p_t x_t} \geq \sup_{0<\xi<\lambda x_t} \frac{(p_{t+1}((f(x_t)-f(x_t-\xi))-p_t\xi))}{(p_t\xi)(p_{t+1}(f(x_t)-f(x_t-\xi)))} \geq M_{p^t,B_t}(S_t)$$

with S_t and B_t defined as before.

Now we may apply Theorem 4.9. Let $(x_t,y_t,c_t)_{t\in N}$ be a program satisfying Condition S such that $\Sigma_{t=1}^{\infty}(\mu(x_t)/p_t x_t)$ is convergent and $\inf p_t x_t>0$. Then $\Sigma_{t=1}^{\infty}M_{p^t,B_t}(S_t)$ is convergent, and since $x_t\leq b(S_t)=-\inf pr_1(S_t)$, each t, we have $a=\inf p_t b(S_t)\geq\inf p_t x_t>0$. Thus, Theorem 4.14 applies and the program is inefficient. Conversely, if $(x_t,y_t,c_t)_{t\in N}$ is an inefficient program satisfying Condition S, then $\inf p_t x_t>0$, and by Theorem 4.12, $\Sigma_{t=1}^{\infty}m_{p^t,B_t}(S_t)$ converges. The same must then hold for $\Sigma_{t=1}^{\infty}(\mu(x_t)/p_t x_t)$.

4.5. MEASURES OF CURVATURE IN THE CONSUMPTION MODEL.

As was shown in Chapter 2, deriving the reduced model from a given consumption model involves, on the one hand, an aggregation or summation over consumers on each single market, on the other hand, a translation of (aggregate) prefered sets to the origin. At present we leave out considerations of aggregation, thus essentially confining ourselves to consumption models with only one consumer in each generation. Since in this case, the operation leading to a reduced model is only a translation, the measures of curvature are easily interpreted as pertaining to preferred sets, or more precisely, to the curvature of the indifference curve at the point in question.

Actually, the hyperbola curvature can be given an interpretation as defining an "ideal" consumer with whom every other consumer is compared.

Suppose that a consumer has the utility function $u:R_+ \longrightarrow R$ defined by $u(x_1,x_2)=ax_1x_2$. Let $x^o=(x_1^o,x_2^o)$ be a consumption bundle with $x_1^o x_2^o=1$ and consider the set $\{x\varepsilon R^2 \,|\, u(x)\geq u(x^o)\}$. Then for $z\varepsilon R^2$ such that $u(x^o+z)\geq u(x^o)$ we have

$$a(x_1^o+z_1)(x_2^o+z_2) \geq ax_1^o x_2^o$$

or

$$z_1 z_2 \geq -x_2^o z_1 - x_1^o z_2$$

If we take as support at (x_1^o,x_2^o) the prices $(u_1'(x_1^o),u_2'(x_2^o))=(ax_2^o,ax_1^o)$ and use that $x_1^o x_2^o=1$, we finally get that

$$p_1z_1+p_2z_2 \geq -a(p_1z_1)(p_2z_2).$$

Thus $S_p(a)$ is the upper level set of the positively homogeneous utility function $u(x)=ax_1x_2$ at any point (x_1^o,x_2^o) with $x_1^o x_2^o=1$. In other words, the curvature is constant along the indifference curve $\{(x_1,x_2)\,|\,x_1x_2=1\}$.

The idea of identifying curvature with degree of similarity with respect to an "ideal" consumer should not be pressed too far, relying as it does heavily on the somewhat doubtful "ideality" of the Cobb-Douglas utility function. On a closer scrutiny the special position of this functional form may be explained by the fact that it is used in empirical work because it is tractable and by theorists because it is used in empirical work.

Anyway, there are other candidates for curvature measure in the case considered, the most prominent one - as well from the point of view of the literature (cf. Balasko and Shell [1980]) as from sheer geometric considerations - would be the Gaussian curvature (see e.g. Laugwitz [1965]) which in the present two-dimensional set-up amounts to approximation by circles.

As was done in the previous section, measures of curvature defined in this way can be evaluated in terms of hyperbolic measures of curvature. It turns out to be easier to do this evaluation in terms of the parabolic curvature discussed in Section 4.4. Thus it can be shown that if $r \leq 1/2b$, then $B_p(r)+R_+^2 \subset S_p^{\perp}(b)+R_+^2$, where $B(r)$ is the disk with center in $r\|p\|$ and radius r, and $p=(p_1,p_2) \in \dot{R}_+ \times \dot{R}_+$. Also, if

$$r \geq \frac{1}{2b}\left(1+ \frac{1}{4}\left(\min\left\{\frac{p_1}{p_2}, \frac{p_2}{p_1}\right\}\right)^2\right)$$

then $S_p(b)+R_+^2 \subset B_p(r)+R_+^2$. Details are left to the reader.

Thus, if in a given consumption model the Gaussian curvatures belong to a compact interval not containing 0, and the ratios p_t/p_{t+1} and p_{t+1}/p_t have upper and lower bounds, then we can draw on the previous results and get that a program is efficient if and only if $\Sigma_{t=1}^{\infty} 1/p_t$ is divergent.

4.6.* OPTIMALITY CRITERIA FOR THE DISCOUNTED UTILITY MODEL.

We return here to the model treated in Sections 2.5 and 3.5. For this model we have a characterization of σ-efficiency which, however, is little operational due to the occurrence of the composition * in the expressions.

At this point we might proceed in a way paralleling our treatment of efficiency, that is, (1) deriving parametric σ-efficiency criteria for subclasses of models, and (2) approximating the remaining models with those in the subclass. There is, however, a shorter way, exploiting the results obtained previously.

THEOREM 4.18. Let $\Sigma=(S_t)_{t\in N}$ be a reduced model with support $(p_t)_{t\in N}$. Then Σ is σ-efficient if and only if

$$\liminf_{t\to\infty} p_t b_t = 0$$

where $b_t = -\inf\{x \mid (x,y)\in S_t\}$.

Proof: We start with the case where $p_t=1$, all $t\in N$. If $\liminf_{t\to\infty} b_t=0$, then, by Example 3.6, the reduced model $\Sigma'=(S_t)_{t\in N}$ with

$$S_t=\{(x,y) \mid x+y \geq 0,\ x\geq -b_t\}$$

is efficient. Clearly, $S_t\subset\Sigma'$, each $t\in N$.

Suppose that Σ is not σ-efficient, and let $(\xi_t)_{t\in N}$ be a σ-improvement. Then $(\xi_t^+)_{t\in N}$, where $\xi_t^+=\max\{0,\xi_t\}$ for each t, is an improvement, since $(-\xi_t,\xi_{t+1})\in S_t$ implies $(-\xi_t^+,\xi_{t+1}^+)\in S_t$ and $\xi_t^+\geq 0$, all t, $\xi_t^+>0$, some t. But according to Lemma 4.6, Σ is efficient, and we have a contradiction.

Conversely, suppose that Σ is not σ-efficient. Then $\inf pr_1(S_1 * \ldots * S_t)$ goes to zero as t goes to infinity. Since for each t we have $S_t\subset S_1*\ldots*S_t$ (for each $(x,y)\in S_t$, we have $(0,0)\in S_\tau$, $\tau>t$, and get $(x+0,y)\in S_1*\ldots*S_t$), we immediately have that $\liminf b_t=0$.

For the case of an arbitrary support $(p_t)_{t\in N}$ we note that the morphism $G=(g_t)_{t\in N}$ with $g_t(x)=p_t x$, $x\in R$, takes Σ to a reduced model $G(\Sigma)=\Sigma'=(S')_{t\in N}$ with support $(1,1,\ldots)$. Also, we note that if $(\xi_t)_{t\in N}$ is a σ-improvement for Σ, then $(p_t\xi_t)_{t\in N}$ satisfies the conditions of the corollary of Theorem 3.16, and therefore $G(\Sigma)$ is not σ-efficient. Now we may apply the results derived above to $G(\Sigma)$ to yield the conclusions of the theorem.●

The characterization of discounted utility maximizing programs obtained in Theorem 4.18 is a special case of the results by Weitzmann [1973], Peleg [1970].

4.7.* DIVERGENT BIRTH PROCESSES.

As a final example of applications of (weak) reduced models we give a brief treatment of the birth process introduced in Section 2.7.

Recall that a birth process was a family $P=(P_n(\cdot))_{n\in N}$ of solutions to the system of equalities (2.6)-(2.7) for given parameters $(\lambda_n)_{n\in N}$. We say that P is divergent if $\Sigma_{n=1}^{\infty}P_n(t)<1$ for some $t>0$. In Section 2.7 we noted that P could be transformed to a weak reduced model $\Sigma=(S_t)_{t\in N}$ where

$$S_n=\{(x,y)\,|\,x \geq P_n(y)-1\},$$

each n, and P divergent meant that there is a sequence $(\xi_n)_{n\in N}$ and $t>0$ such that $\xi_n\geq 0$, $(-\xi_n,t)\in S_n$, each n, and $\Sigma_{n=1}^{\infty}(-\xi_n+1)-1>0$.

In view of the above, it would seem appropriate to introduce a composition, $\bullet$, yielding sets

$$S_1\bullet S_2=\{(x,y)\,|\,\exists x_i,i=1,2:(x_i,y)\in S_i,i=1,2,x\geq(x_1+1)+(x_2+1)-1\},$$

and, in general,

$$S_1\bullet\ldots\bullet S_n=$$
$$=\{(x,y)\,|\,\exists x_i,i=1,\ldots,n:(x_i,y)\in S_i,i=1,\ldots,n,x\geq\Sigma_{i=1}^{n}(x_i+1)-1\}.$$

Note that, since from (2.6) we get

$$\frac{d}{dt}\,(\Sigma_{i=1}^{n}P_i(t))=-\lambda_nP_n(t)\,\leq 0$$

we have that $S_1\bullet\ldots\bullet S_n$ satisfies the conditions (Ω1) and (Ω2) of Section 3.1 and thus the properties demanded of the constituent sets of a weak reduced model.(cf. Definition 2.12)

Until now, the analysis has been analoguous to that of the preceding sections. However, divergence of P cannot be decided upon from the behaviour of inf $pr_1(S_1\bullet\ldots\bullet S_n)$, since for every birth process we will have inf $pr_1(S_1\bullet\ldots\bullet S_n)=-1$, all n. This is a consequence of the following lemma:

LEMMA 4.19. Let $P=(P_n(\cdot))_{n\in N}$ be a birth process. Then for each n,

$$\Sigma_{i=1}^{n} P_i(t)\to 0 \text{ for } t\to\infty.$$

<u>Proof</u>: By induction on n. For n=1, solving (2.7) yields $P_1(t)=e^{-\lambda_1 t}$ which goes to 0 as $t\to\infty$.

Suppose that $n\geq 1$ and $\Sigma_{i=1}^{n-1}P_i(t)\to 0$ for $t\to\infty$. If $\Sigma_{i=1}^{n}P_i(t)$ does not go to zero, then it must tend to some $r>0$. Choose ξ with $r>\xi>0$ annd T so large that $\Sigma_{i=1}^{n-1}P_i(t)<\xi$, $r\leq\Sigma_{i=1}^{n}P_i(t)<r+\xi$ for all $t\geq T$. Then $P_i(t)>r-\xi>0$ for all $t\geq T$ contradicting the fact that

$$\frac{d}{dt}\ \Sigma_{i=1}^{n}P_i(t)=-\lambda_n P_n(t)\to 0$$

for $t\to\infty$. We conclude that $\Sigma_{i=1}^{n}P_i(t)\to 0$ as $t\to\infty$.●

In order to proceed in a way similar to that of the other applications considered, we must modify slightly the construction of the reduced model: For t>0, define

$$S_n(t)=\{(x,y)\mid x\geq P_n(y)-1,\ y\leq t,\ x\geq P_n(t)-1,\ y\geq t\}.$$

Then we have the following characterization of divergence, the proof of which is trivial:

THEOREM 4.20. Let $P=(P_n(\cdot))_{n\in N}$ be a birth process. Then P is divergent if and only if

$$\sup_{n\geq 1}\inf\ pr_1(S_1(t)\bullet\ldots\bullet S_n(t)) < 0$$

for some t>0.●

Our final goal is to characterize divergent birth processes in terms of the parameters $(\lambda_n)_{n\in N}$. It seems, however, to be rather tedious to establish a counterpart of Theorem 3.12, so we shall take a direct approach:

THEOREM 4.21. Let $P=(P_n(\cdot))_{n\in N}$ be a birth process with parameters $(\lambda_n)_{n\in N}$. Then P is divergent if and only if $\Sigma_{i=1}^{\infty}1/\lambda_i$ converges.

<u>Proof</u>: "If": Choose $\bar{n}\in N$, $\bar{n}>1$, such that $\lambda_1\Sigma_{i=\bar{n}+1}^{\infty}1/\lambda_i<1/4$. By Lemma 4.19, there is T such that

$$\inf pr_1(S_1(T)\bullet\ldots\bullet S_n(T)) \leq -3/4.$$

Now, from (2.6) we have that for each $n>1$, $P_n(t)$ maximal implies $P_n'(t)=0$ or

$$-\lambda_n P_n(t)+\lambda_{n-1}P_{n-1}(t)=0,$$

from which we get that

$$\max_{t>0} P_n(t) \leq \frac{\lambda_{n-1}}{\lambda_n} \max_{t>0} P_{n-1}(t),$$

yielding $P_n(t) \leq \lambda_1/\lambda_n$ for all $n>1$. Consequently, for each $n>\bar{n}$ we have

$$\inf pr_1(S_1(T)\bullet\ldots\bullet S_n(T)) \leq (-\tfrac{3}{4}+1)+\Sigma_{i=\bar{n}+1}^{n}P_i(t)-1 \leq$$
$$\leq \tfrac{1}{4}+\Sigma_{i=n+1}^{\infty}\frac{\lambda_1}{\lambda_i}-1=-\tfrac{1}{2}$$

and P is divergent by Theorem 4.20.

"Only if": Suppose that P is divergent and let $r>0$ be a number such that

$$\inf pr_1(S_1(T)\bullet\ldots\bullet S_n(T)) < -r$$

for all n. Then $\Sigma_{i=1}^{n}P_i(T)<1-r$, and since $\Sigma_{i=1}^{n}P_i(T)=1-\int_0^T\lambda_n P_n(t)dt$, we get that $\int_0^T P_n(t)dt>r/\lambda_n$. Summing over n and using $\Sigma_{i=1}^{n}P_i(t)\leq 1$ we get

$$T \geq \int_0^T\Sigma_{i=1}^{n}P_i(t)dt > r\Sigma_{i=1}^{n}\frac{1}{\lambda_i}$$

for all n, thus $\Sigma_{i=1}^{\infty}1/\lambda_i$ must be convergent.●

As mentioned in Section 2.7, the result of Theorem 4.21 is not new - it can be found with a shorter proof in Feller [1968]. However, as the divergence criterion displays a spectacular similarity to the efficiency criteria considered in this work, it would seem that the problems are related, and that they should allow for treatment by similar methods. What has come out of this section is that in broad outline, this is indeed the case. Also, the proof given above may yield some further insight into the problem.

CHAPTER 5: APPROXIMATING SETS AND MEASURES OF CURVATURE

In the preceding chapter we introduced the family of sets $S(a;p_1,p_2)$ and showed how they could be used to approximate a reduced model from the inside and from the outside so as to get an efficiency criterion by approximation.

At several instances we argued that this family of sets has several attractive properties. Firstly, they are easy to work with since the composition o on sets in Ω^1 translates to addition of parameters (with correction for price changes). Secondly,many of the existing efficiency criteria in the literature can be derived from criteria obtained in this way, and thirdly, the sets $S(a;p_1,p_2)$ have a clear economic interpretation in at least one of the models giving rise to efficiency considerations (the consumption model introduced in Chapter 1).

Partly guided by the results mentioned we introduce, in Section 5.1, four axioms which seem desirable for a family of approximating sets. We then show, in Section 5.2, that these axioms completely determine the family as Ω^1, the family whose members are given by

$$S(a;p_1,p_2)=\{(x,y)\,|\,p_1x+p_2y\geq 0,\quad p_1x+p_2y+a(p_1x)(p_2y)\geq 0\}$$

for $a\varepsilon R_+$, $p_1,p_2\varepsilon \dot{R}_+$ with $S(\infty;p_1,p_2)=R_+^2$ for $p_1,p_2\varepsilon \dot{R}_+$.

5.1. AXIOMS FOR AN APPROXIMATING FAMILY

We list the axioms in full generality. In Section 5.3 we will use them, to begin with , for the case where all prices are equal to 1 to determine the corresponding subfamily satisfying Axioms I, II and IV. Then Axiom III is used to establish that the complete family is Ω^1.

Let a mapping, $Z: \overline{R}_+ x\dot{R}_+ x\dot{R}_+ \longrightarrow \Omega$,where $\overline{R}_+ = R_+ \cup \{\infty\}$, be given; Z takes triples $(a;p_1,p_2)$ into sets $Z(a;p_1,p_2)\varepsilon\Omega$. In the interpretation a is a measure of curvature (cf. the discussion in Sections 4.4 and 4.5). Clearly a perfectly arbitrary map Z will not give an interesting measure of curvature. Also, as we have seen in previous chapters,many different - often conflicting - views are possible as to what should constitute a "good measure" of curvature.

In such circumstances, it is customary to use an axiomatic approach; to state as axioms certain properties which a "good" measure of curvature should have, and then to derive from those axioms some - hopefully small - subclass of all conceivable measures of curvature such that these properties are satisfied.

Below we state a set of axioms for measures of curvature. We shall comment upon these axioms in the following section.

AXIOM I: For $a,b\varepsilon\overline{R}_+$, $p_1,p_2,p_3\varepsilon\dot{R}_+$,

$$Z(a;p_1,p_2)oZ(b;p_2,p_3)=Z(a+b;p_1,p_3)$$

Here o is the composition introduced in Chapter 3.

AXIOM II: For $\lambda\varepsilon\dot{R}_+$, $(a;p_1,p_2)\varepsilon\overline{R}_+ x\dot{R}_+ x\dot{R}_+$

$$\lambda Z(a;p_1,p_2)=Z(\frac{a}{\lambda};p_1,p_2)$$

AXIOM III: For $\mu_1,\mu_2\varepsilon\dot{R}_+$, $(a;p_1,p_2)\varepsilon\overline{R}_+ x\dot{R}_+ x\dot{R}_+$

$$Z(a;\mu_1 p_1,\mu_2 p_2)=\{(z_1,z_2)|z_1=\frac{x_1}{\mu_1} , z_2=\frac{x_2}{\mu_2} , (x_1,x_2)\varepsilon Z(a;p_1,p_2)\}$$

AXIOM IV: $Z(1;1,1)$ is symmetric around the diagonal, that is, $(x,y)\varepsilon Z(1;1,1)$ if and only if $(y,x)\varepsilon Z(1;1,1)$. Furthermore, there is a

C^2-function $f:R_+ \longrightarrow R_+$ such that

(i) $x=f(y)$ if and only if $(-x,y)\varepsilon bdZ(1;1,1)\cap(-R_+ xR_+)$
(ii) $f'(0^+)=1$ and $f''(0^+)=-2$

Let Ω^o denote the image of any mapping Z satisfying Axioms I-IV.

5.2. DISCUSSION OF THE AXIOMS

Below we comment in some detail on the axioms of Section 5.1.

Axiom III is an axiom of invariance of the measure a for different choices of units for the commodities. Let (x,y) belong to the boundary of $Z(a;p_1,p_2)$ with $x<0$. A reduction of $|x|$ units of the first good may be compensated by an increase of y units in the second good. Changing the units of measurement, so that the new units of the first and second good are μ_1 and μ_2 respectively of the old units , implies that $|x|/\mu_1$ new units of the first good may be compensated by y/μ_2 new units of the second good. The prices of the new units will be $\mu_1 p_1$ and $\mu_2 p_2$ where p_1 and p_2 are the original prices.

Since the choice of units is arbitrary it is desirable that the measure of curvature should be invariant under changes in units. Let

$$V(a;p_1,p_2)= \{(v_1,v_2)|(v_1,v_2)=(p_1x_1,p_2x_2),\ (x_1,x_2)\varepsilon Z(a;p_1,p_2)\}$$

These are the pairs of values corresponding to $Z(a;p_1,p_2)$. By Axiom III; $V(a;p_1,p_2)=V(a;1,1)$ and, by construction , $V(a;1,1)=Z(a;1,1)$. Hence the set $V(a;p_1,p_2)$ is, for fixed a, independent of the prices.

Note that one consequence of Axiom III is that for $a\varepsilon\overline{R}_+$ and $\lambda\varepsilon\dot{R}_+$, $Z(a;1/\lambda,1/\lambda)=\lambda Z(a;1,1)$ and thus this latter set belongs to Ω^o.

Axiom II implies that the sets have been indexed so that $\lambda Z(a;1,1)=Z(a/\lambda;1,1)$. There is another, related, reason for demanding that $\lambda Z(a;p_1,p_2)\varepsilon\Omega^o$ if $Z(a;p_1,p_2)\varepsilon\Omega^o$. To see this consider a λ-replica economy of the overlapping generations model; λ a natural number. If the reduced model corresponding to the original model is $(S_t)_{t\varepsilon N}$ then the reduced model associated with the replica economy will be $(\lambda S_t)_{t\varepsilon N}$

and Axiom III ensures that $S_t \varepsilon \Omega^o$ if and only if $\lambda S_t \varepsilon \Omega^o$. Again Axiom II implies a relation between the measure of curvature for S_t and λS_t if they happen to belong to Ω^o.

Axiom I asserts that the composition in Ω^o corresponds to addition of parameters. The simplifications that result from such a simple relation between the set theoretic operation of composition and the algebraic operation of addition are evident from the discussion in Chapter 3.

Axiom IV, finally, is just a regularity axiom. We are only interested in the sets $Z(a;p_1,p_2)\cap(-R_+xR_+)$ so the assumption of symmetry is made solely in order to simplify notation. The choice of the value 1 for $f'(0^+)$ is dictated by the requirement that (1,1) should be a support for $Z(a;1,1)$ and the choice $f''(0^+)=-2$ is a normalization. It singles out, among the sets $Z(a;1,1)$, the one that should have index 1.

A mapping Z, satisfying Axioms I-IV is not a parametrization since it fails to be injective. However, by introducing a suitable equivalence relation, and defining the associated induced mapping this can be remedied (cf. Remark 3.13).

5.3. THE APPROXIMATING FAMILY DETERMINED BY THE AXIOMS

We will now show that Z satisfies the Axioms I-IV if and only if $Z(a;p_1,p_2)=S(a;p_1,p_2)$. It is convenient to prove this in three steps.

THEOREM 5.1. Let $\bar{Z}$ be the restriction of Z to $\dot{R}_+x\{1\}x\{1\}$. Then $\bar{Z}$ satisfies Axioms I, II and IV if and only if

$$\bar{Z}(a;1,1)=S(a;1,1)$$

for $a\varepsilon R_+$.

Before giving the proof of Theorem 5.1 we also state

LEMMA 5.2. Let $Q:R_+x\dot{R}_+x\dot{R}_+\longrightarrow\Omega$ be a mapping such that $Q(a;1,1)=S(a;1,1)$ for $a\varepsilon R_+$ and such that Q satisfies Axiom III. Then

$$Q(a;p_1,p_2)=S(a;p_1,p_2).$$

Combining Theorem 5.1 and Lemma 5.2 we have

THEOREM 5.3. $\Omega^o=\Omega^1$.

<u>Proof of Theorem 5.1</u>:The "if" part is straightforward and is left to the reader. For the "only if" part, let x, y, z, a and $\bar{a}$ denote positive reals unless otherwise stated. Our aim is to derive a differential equation for the function f defined by $x=f(y)$ if and only if $(-x,y)\epsilon \mathrm{bd}Z(1;1,1)\cap(-R_+\cap R_+)$ (as in Axiom IV(i)).

(a) $(-x,y)\epsilon \mathrm{bd}Z(a;1,1)$ if and only if $x=a^{-1}f(ay)$

Axioms II, III and IV give the following chain of equivalences:

$$(-x,y)\epsilon \mathrm{bd}Z(a;1,1) \Leftrightarrow (-ax,ay)\epsilon \mathrm{bd}Z(1;1,1) \Leftrightarrow ax=f(ay) \Leftrightarrow x=a^{-1}f(ay)$$

Define $f_a(y)=a^{-1}f(ay)$ and $f_0(y)=y$. Note that $f_a(\cdot)$ is non-decreasing since $Z(1;1,1)$ is strictly monotone.

(b) The map $a\rightarrow f_a(y)$ is C^1 and, in particular,

$$\lim_{a\to 0}\frac{f_a(y)-f_0(y)}{a}=\lim_{\bar{a}\to 0}\frac{df_a(y)}{da}\bigg|_{a=\bar{a}}=-y^2$$

Using that f is C^2, $f'(0^+)=1$, $f''(0^+)=-2$ and expanding in Taylor series

$$\lim_{a\to 0}\frac{f_a(y)-f_0(y)}{a}=\lim_{a\to 0}\frac{f(ay)-ay}{a^2}=$$

$$=y^2 \lim (ay)^{-2}[f'(0^+)(ay)+\tfrac{1}{2}f''(0^+)(ay)^2+o((ay)^2)-ay]=-y^2$$

On the other hand, we have

$$\frac{df_a(y)}{da}\bigg|_{a=\bar{a}}=\frac{\bar{a}f'(\bar{a}y)-f(\bar{a}y)}{\bar{a}^2}=y^2[(ay)^{-2}((\bar{a}y)f'(ay)-f(\bar{a}y))]$$

Expanding f and f' in Taylor series we get

$$\lim_{\bar{a}\to 0} \frac{df_a(y)}{da}\bigg|_{a=\bar{a}} = -y^2$$

(c) $$\frac{d}{da} f(f_a(y))\bigg|_{a=0} = -f'(y)y^2$$

The map $a \to f(f_a(y))$ is C^1 and the derivative in (c) is

$$\frac{d}{da} f(f_a(y)) = f'(f_a(y))\frac{d}{da} f_a(y)$$

evaluated at 0.

(d) $f_a(f_b(y)) = f_{a+b}(y)$

Let $(-x,y)\varepsilon Z(a+b,1,1) = Z(a,1,1)oZ(b,1,1)$. Then there is z such that $z \leq f_b(y)$, $x \leq f_a(z)$ and, since $f_a(\cdot)$ is non-decreasing, $z \leq f_a(z) \leq f_a(f_b(y))$. For a given y, there exists x such that equality is achieved; choose $z = f_b(y)$ and $x = f_a(z)$. Hence every boundary point of $Z(a+b;1,1)$ satisfies $x = f_a(f_b(y))$ as well as $x = f_{a+b}(y)$ which proves assertion (d).

(e) $$\frac{d}{da} f(f_a(y))\bigg|_{a=0} = yf'(y) - f(y)$$

Since $f(f_a(y)) = f_{1+a}(y) = (1+a)^{-1} f((1+a)y)$, we get

$$\frac{d}{da} f(f_a(y) = \frac{(1+a)yf'((1+a)y) - f((1+a)y)}{(1+a)^2}$$

which evaluated at $a=0$ gives the expression in (e).

Combining (c) and (e), it is seen that f satisfies the following differential equation for $y \geq 0$

$$-f'(y)y^2 = yf'(y) - f(y)$$

This equation has solutions $f(y)= K(y(1+y)^{-1})$, K a constant, which is C^2 on R_+. From $f'(0^+)=1$ we get K=1.

We conclude that if $(-x,y)\varepsilon bdZ(1;1,1)$, then $x=(y(1+y)^{-1})$ or $x+y+xy=0$. Since Z(1;1,1) is strictly monotone and symmetric, we get that Z(1;1,1)=S(1;1,1). The conclusion now follows from a further application of Axiom II. ●

Proof of Lemma 5.2: By Axiom III

$$Q(a;p_1,p_2) = \{(z_1,z_2) \mid z_1=x_1/p_1,\ z_2=x_2/p_2,\ (x_1,x_2)\varepsilon Q(a;1,1)\} =$$

$$= \{(z_1,z_2) \mid z_1=x_1/p_1,\ z_2=x_2/p_2,\ (x_1,x_2)\varepsilon S(a;1,1)\} = S(a;p_1,p_2)$$

which proves the assertion.●

Proof of Theorem 5.3: It remains to show that $Z(\infty,1,1)$ is R^2_+ and that Z(0;1,1) is the closed upper half space determined by the vector (1,1). Let H be this upper half space.

Since $Z(a,1,1)\subset H$ for $a\geq 0$, we get, using Axioms I, II, and the definition of o

$$Z(a;1,1) = Z(a;1,1)oZ(0;1,1)\subset HoZ(0;1,1)=Z(0;1,1)$$

for $a\geq 0$. Hence Z(0;1,1) contains $\cup_{a>0}Z(a;1,1)=\text{int}H\cup\{0\}$ and, since Z(0;1,1) is closed, Z(0;1,1)=H.

By assumption $R^2_+\subset Z(\infty;1,1)$ and we have

$$Z(\infty;1,1)=Z(a;1,1)oZ(\infty;1,1)\subset Z(a;1,1)$$

which implies

$$R^2_+\subset Z(\infty;1,1)\subset[\cap_{a\geq 0}Z(a;1,1)]=R^2_+ .\bullet$$

5.4. A FURTHER PROPERTY OF THE FAMILY Ω^o

The results in Sections 5.1-5.3 show that the family Ω^o has some nice properties. Going back to the construction of a reduced model from a consumption model and/or a production model one problem appears to have been glossed over; namely the relation between the curvature of individual preferred sets and/or production sets and the curvature of their sum.

When the original model has more than one new agent appearing at each date then we would like to relate the curvature or approximation of each S_t in the reduced model to the curvature of the individual preferred sets for consumers and production sets for producers. When these sets happen to belong to Ω^o there is a simple relation between the measures of curvature. The following example shows that this can not be expected for arbitrary preferred sets or production sets.

EXAMPLE 5.4. Let Z^1 and Z^2 be defined as follows:

$$Z^1=S(1;1,1)\cap\{(x,y)\,|\,y\geq 0\}$$

$$Z^2=S(1;1,1)\cap\{(x,y)\,|\,x\geq 0\}$$

Then the inner curvature of Z^1 and Z^2 are both ∞. On the other hand since $0\in Z^1\cap Z^2$ and $Z^1\cup Z^2$ equals $S(1;1,1)$ we get

$$S(1;1,1)\subset Z^1+Z^2$$

To see that we in fact have equality in the relation above note first that Z^1+Z^2 is closed. Let p be any support to $S(1;1,1)$. Then since $Z^i\subset S(1;1,1)$ for $i=1,2$, p is a support to Z^i, $i=1,2$ and consequently to Z^1+Z^2. As this is true for any support, Z^1+Z^2 can have no point in the exterior of $S(1;1,1)$ and, since $S(1;1,1)$ as well as Z^1+Z^2 are closed, $Z^1+Z^2=S(1;1,1)$.

The largest member of Ω^o contained in $Z^i, i=1,2$, is R^2_+ and since $R^2_++R^2_+=R^2_+$ the sum of these approximations is not a good approximation of the sum Z^1+Z^2.●

Thus, if the set S_t is constructed from (the translations to 0) of individual preferred sets, then the inner curvature of S_t will, in general, be smaller than the sum of inner curvatures of the preferred sets. For the outer curvature an analogous inequality holds; the outer curvature of a sum of sets will, in general, be larger than the sum of outer curvatures.

Consider now the case where S_t is derived from individual preferred sets all belonging to Ω^1. The subfamilies of Ω^1, Ω', corresponding to some fixed prices p_1, p_2, have the property that there is a set $A\varepsilon\Omega'$ such that

$$[B\varepsilon\Omega' \text{ and } B \text{ is not a hyperplane or } R_+^2\ ] \Rightarrow B=\alpha A \text{ for some } \alpha>0.$$

The following theorem shows that when the approximating family has this property then, with a suitable numbering of the sets of the family, there will be a simple relation between the curvatures of individual preferred sets and the curvature of their sum. We need a (well-known) result on sums of convex sets

LEMMA 5.5. Let A be a convex set . Then for $\alpha,\beta\varepsilon R_+$

$$\alpha A+\beta A=(\alpha+\beta)A$$

Proof: If α or β (or both) are 0 the relation is trivial. For α and β positive we have

$$\frac{\alpha}{\alpha+\beta}\ A+\frac{\beta}{\alpha+\beta}\ A=A$$

by convexity of A, so

$$\alpha A+\beta A=\frac{\alpha}{\alpha+\beta}\ (\alpha+\beta)A+\frac{\beta}{\alpha+\beta}\ (\alpha+\beta)A=(\alpha+\beta)A. \quad \bullet$$

THEOREM 5.6. Let $Z:\overline{R}_+ x\dot{R}_+ x\dot{R}_+ \longrightarrow\Omega^0$ be an approximating family satisfying Axiom II. Then

$$\Sigma_{i=1}^m Z(a_i;p_1,p_2)=Z((\Sigma_{i=1}^m\ \frac{1}{a_i})^{-1};p_1,p_2)$$

Proof: Suppose that a_i is positive for i=1,...,m. Using Axiom II and Lemma 5.5, we get that

$$\Sigma_{i=1}^m Z(a_i;p_1,p_2)=\Sigma_{i=1}^m\ \frac{1}{a_i}Z(1;p_1,p_2)=$$

$$=(\Sigma_{i=1}^m\ \frac{1}{a_i})Z(1;p_1,p_2)=Z((\Sigma_{i=1}^m\ \frac{1}{a_i})^{-1};p_1,p_2)$$

We leave it to the reader to verify the theorem for the situation where $a_i=0$ or $a_i=\infty$ for some i. $\bullet$

PART II

MANY-GOODS MODELS

CHAPTER 6: MARKET STRUCTURES AND OVERLAPPING MARKETS ECONOMIES

In Part I we have developed a method for analyzing efficiency of infinite horizon models. Since the emphasis was on new concepts and tools, we restricted the treatment to one-good models, so that in the models considered only a single good was available at every date.

The principal aim of Part II is to provide a generalization of the methods and results of Part I to models where at each date an arbitrary finite number of commodities are available for consumption or production, depending on the particular model to be studied . A large part of this generalization poses little or no problems. Consequently, we shall be rather brief on problems which were treated in detail in Part I and display no new features in the general case. However, some aspects must to be treated with a certain care if the results are to carry over, and we shall comment on this in due course.

As a starting point for the treatment of the general case, it will prove useful to devote some space to a detailed study of the commodity space underlying the general infinite-horizon models. This is the topic of the present chapter which has no counterpart in Part I. The following chapters, then, will proceed to generalize the main results of Chapters 2-5 to the many-goods case.

6.1 MARKET STRUCTURES

A crucial part of the infinite horizon production model was the technology assumption that production transforms input at one date to output at the next date. Similiarly, in the consumption model each generation lives in two periods, and a member of any generation may exchange goods with the older generation in the first period and with the younger generation in the second.

What matters for the model is of course not the picturesque stories about life and death supplied in order that the reader may visualize the way in which the model works, but the possibilities of exchange linking production at different dates or consumption for different generations. This is captured by the concept of market structure.

DEFINITION 6.1. Let C be a non-empty set. A <u>market structure on C</u> is a family M of non-empty, finite subsets of C such that

(i) $\cup_{m\in M} m = C$

(ii) for each $c\in C$, $M(c)=\{m\in M \mid c\in m\}$ is finite.

Let M be a market structure on C and M' a subset of M. M' is <u>connected</u>, if for any two non-empty subsets M_1, M_2 of M' such that $M_1\cup M_2=M'$ there are $m_1\in M_1$, $m_2\in M_2$ such that $m_1\cap m_2\neq\phi$. ●

If M is itself connected then the market structure is connected. In the sequel we shall assume, unless there is a statement to the contrary, that any market structure is connected. It will be evident from the discussion to follow that the study of market structures which are not connected can be reduced to that of connected market structures.

The set C may be considered as a set of names or labels of commodities, where in accordance with standard practice in general equilibrium theory (cf.Debreu [1959],ch.2) commodities may be distinguished by quality, location, date and event upon which delivery is contingent. The subsets of C belonging to the family M are interpreted as markets open to some agents, to be introduced later, in the sense that if m is the market for agent i and $c_1, c_2 \in m$, then i can trade commodity c_1 against commodity c_2 at some prices.

Condition (i) of the definition says that every commodity should belong to some market and is no real restriction. If some c was in no m, then this c would be of no interest in our theory, as it could at most be dispensed with but never acquired, and it could be deleted from C. Note that (i) does not really say that every commodity can be traded against some other commodity since $c \in C$ may belong only to $\{c\}$, but this rather special case, and, more generally, cases of one-commodity markets which have no meaningful interpretation, will pose no problems as they will be either excluded by other assumptions or otherwise play no role for the theory.

Condition (ii) plays a central role as it allows us to aggregate over markets: If at each market m containing a commodity c a net trade z_c^m of commodity c has been performed, then the total net trade can be found as $\Sigma_{m \in M(c)} z_c^m$. It should be noticed that (ii) is not the only finiteness assumption made since we have demanded also that each $m \in M$ is a finite set. Furthermore, a third finiteness assumption will follow when we introduce economies over a market structure. All these finiteness assumptions serve the same purpose: to make it possible to use some simple linear operations as commodity-wise addition, scalar products on each market and so on, essentially as in finite economies.

A market structure M on a set C bears a certain resemblance with a topology but clearly M is not a topology except in very special cases. Finite unions of sets m in M are finite sets, but the set of all finite unions of sets in M does not satisfy (ii) which is central for the following theory. Therefore we shall not use general topology for the analysis of market structure. However, many of the concepts which are useful have close counterparts in topology. This goes, in particular, for connectedness which says that the market structure cannot be split into two market structures with no mutual overlapping of markets.

The following are some examples of market structures:

EXAMPLE 6.2. The Trivial Market Structure. Let C be a finite set, and $M = \{C\}$. This market structure is of course particularly simple, and, as we shall see, no new theory need to be written for economies where the market structure is $\{C\}$. ●

EXAMPLE 6.3. The Discrete Market Structure. Let C be a finite set and let M={all non-empty subsets of C}.Although this market structure differs from that of the preceding example,for the theory of this book it does not pose any new problems compared with the preceding one and both are of minor interest as C is finite. ●

EXAMPLE 6.4. The Market Structure of a Simple Consumption Model. Let $C=N\cup\{0\}$ and let

$$M=\{\{0,1\},\ \{1,2\},\ \ldots,\ \{t,t+1\},\ \ldots\}.$$

Then M is a market structure on $N\cup\{0\}$. We identify each market with the life span of the generation born in its first period. Note that in the interpretation, the possibility for each consumer to trade the commodity in his first period against the commodity in his second implies that there are forward markets, or capital markets, enough to link all time periods with each other. This is actually one of the essential features of overlapping-markets models. ●

EXAMPLE 6.5. The Canonical Market Structure. The example above may be generalized to situations where each generation has an arbitrary finite number of commodities in its associated market. Let $C=N\cup\{0\}$ and let t^n, s^n,r^n,q^n, $n\varepsilon C$, be sequences such that ,$t^o=0$ and , for $n\varepsilon C$,

$$t^n<s^n\leq r^n<q^n$$

$$t^{n+1}=r^n$$

$$s^{n+1}=q^n$$

Let $[a,b]\subset C$ be equal to $\{a,a+1,\ldots,b\}$ for $a\leq b$ and define a market structure by

$$M=\{[t^o,q^o],[t^1,q^1],\ldots,[t^n,q^n],\ldots\}$$

The market structure may be visulized as follows (the first line is relevant only for one of the interpretations to be discussed below):

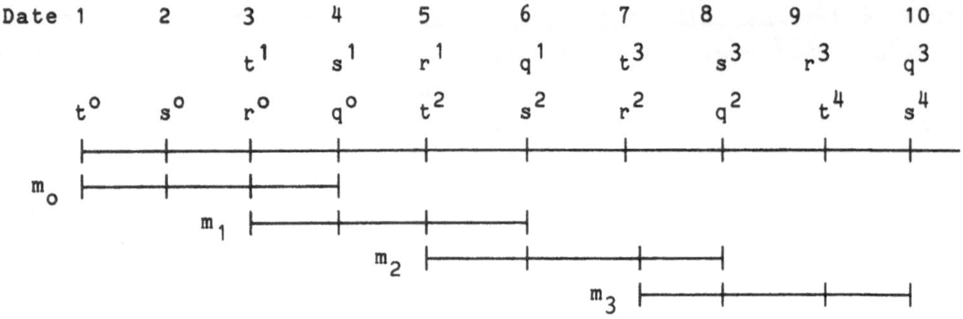

As before we may associate with each market a generation having access to the possibility of exchanging each of the commodities in a market for each other commodity in the same market.

Note that now commodities and lifespans may be interpreted in at least two ways: The generation n may be born at date t^n and live for $q^n - t^n$ periods, consuming one commodity in each period, or it may be born at date 3n and live for three periods, consuming commodities $t^n, \dots s^n - 1$ in the first period, $s^n, \dots, r^n - 1$ in the middle and, in the last period, commodities $r^n, \dots, q^n$.

In this interpretation the commodities $0, \dots, r^o - 1$ can only be exchanged between members of generation 0 and, for each n, commodities $s^n, \dots, r^n - 1$ can only be exchanged between members of generation n. These internal commodities are of less interest to us and we shall often disregard them since they are of no consequence for the theory. ●

In Chapter 4 we used morphisms to extend results valid for a small class of models to a larger class. In this connection morphisms were used only to change the units of measurement of the goods and the market structure was fixed. For the many-goods case we would like morphisms to handle a change in the units of measurement as well as a change in the market structure. This will be accomplished by the concept defined below.

DEFINITION 6.6. Let (C,M) and (C',M') be market structures. A morphism is a pair (η, α) where $\alpha \in R^C_{++}$ and $\eta = (\gamma, \iota)$ such that

(i) $\gamma: C \to C'$ is a map between commodity spaces such that $\gamma^{-1}(m')$ is a finite union of sets from M for each $m' \in M'$.

(ii) $\iota: M \to M'$ is a map assigning to each market in M a market in M' so that $m \subset \gamma^{-1}(\iota(m))$ for $m \in M$.

If (η,α) is a morphism and $\gamma:C \to C'$ is a bijection then (η,α) is a simple morphism. If (η,α) is a morphism and $\alpha=(1,1,1,...)$ then (η,α) is a market morphism. A morphism which is both a simple morphism and a market morphism is a simple market morphism. ●

Let A be a subset of R^C and let 1_A denote the vector defined by

$$1_A(c)=\begin{cases}1 \text{ for } c\varepsilon A \\ 0 \text{ otherwise}\end{cases}$$

We use the convention that $1_c=1_{\{c\}}$ for $c\varepsilon C$

A morphism (η,α) induces a mapping $f_{\eta,\alpha}$ from the basis of $R^{(C)}$, the set of all vectors in R^C with finite support, to $R^{(C')}$ defined by

$$f_{\eta,\alpha}(1_c)=\alpha_c 1_{\gamma(c)}$$

$f_{\eta,\alpha}$ may then be extended by linearity to a mapping $g_{\eta,\alpha}:R^{(C)} \to R^{(C')}$

Essentially, what a market morphism does is that it relabels the commodities and transforms markets, for example, by merging several small markets to a single one. When (η,α) is a morphism, but not a market morphism, (η,α) may transform markets and, through the induced linear mapping $g_{\eta,\alpha}$, change the units of measurement simultaneously.

The need for the map ι in (ii) of Definition 6.6 arises from the fact that there may exists markets $m\varepsilon M$ such that $m\subset\gamma^{-1}(m')$ and $m\subset\gamma^{-1}(n')$ for two different markets $m',n'\varepsilon M'$. ι makes sure that there is a unique market in M' to which such an m is assigned. There will be several examples of morphisms in the sequel.

In the following we write (C,M) for a market structure M on C and $\eta:(C,M) \to (C',M')$ for a market morphism. When (η,α) is not a market morphism, so that $\alpha_c \neq 1_C$, we will in each instance specify α. The reader interested in the formal aspects of the theory may note that the triple $(id_C, id_M, id_R(C))$ where id_C, id_M, and $id_R(C)$ is the identity on C,M and $R^{(C)}$ respectively is a morpism from $(C,M,R^{(C)})$ to itself, and that composition of morphisms will give a new morphism. Thus we have a category of market structures with commodity spaces.

6.2. COUNTABILITY PROPERTIES AND THE CANONICAL MARKET STRUCTURE

Market structures are of interest from the point of view of infinite horizon models only when the set of commodities is infinite. However, market structures do not have arbitrary cardinality. Actually, if (C,M) is a (connected) market structure, then C, and hence M, is countable. This might not be obvious from Definition 6.1 but it is a consequence of the finiteness conditions imposed. As a by-product of the proof of this result, we get a very useful canonical representation of market structures.

THEOREM 6.7. Let (C,M) be a market structure satisfying (i)-(iii) of Definition 6.1. Then

(1) C and M are at most countable

(2) There is a market structure M' on C and a simple market morphism $\eta:(C,M)\longrightarrow(C,M')$, such that M' satisfies

(i) $M'=\{m_0', m_1', \ldots, m_t', \ldots\}$, $m_0'\varepsilon M$

(ii) $m_i'\cap m_j'\neq\phi$ if and only if $j\varepsilon\{i-1,i,i+1\}$, $i,j\varepsilon N\cup\{0\}$.

The market m_0 is <u>the first market</u> in M', and M' is the m_0-<u>canonical form of M</u>.

<u>Proof</u>: Choose $m_0'\varepsilon M$ arbitrarily, and define for $i\varepsilon N$, m_i' inductively by

$$m_i'=\{c\varepsilon C\mid \exists m\varepsilon M\colon c\varepsilon m,\ m\cap m_{i-1}'\neq\phi,\ m\cap(C\setminus\cup_{j=1}^{i-1}m_j')\neq\phi\}$$

and let $M'=\{m_0', m_1', \ldots, m_t', \ldots\}$. Each of the sets m_i' is finite: For m_0' this is obvious; suppose that m_j', $j<i$ are all finite. Then the set of all $m\varepsilon M$ containing some element of $\cup_{j=0}^{i-1}m_j'$ is finite as well, and also

$$\cup M(c) \quad \text{where } c\varepsilon\cup_{j=0}^{i-1}m_j'$$

is finite. Now, m_i' is a finite set since it is a subset of the above set.

Let $C'=\cup_{i=0}^{\infty}m_i'$. If $C'\neq C$, then choose $c\in C\setminus C'$ and $m\in M(c)$. Then $m\cap m_i'=\phi$ for all $i\in N\cup\{0\}$, since otherwise $c\in m_{i+1}'\subset C'$. It follows that $\{m|m\in M(c),c\in C'\}$ and $\{m|m\in M(c),c\in C\setminus C'\}$ partitions M into non-empty sets such that any two markets belonging to each of these sets are disjoint, contradicting connectedness of M. We conclude that $C'=C$.

Now (1) follows immediately since C is covered by a countable collection of finite sets, and $M=\cup_{c\in C}M(c)$.

To prove (2) we need only show that M' satisfies (2.ii) since then M' is a market structure; it is clear from the construction that $\gamma=id_C$ and $\iota:M\longrightarrow M'$ defined by $\iota(m)=m_{i(m)}$ where, for $m\in M$, $i(m)$ is the smallest index such that $m\subset m_i'$, is a simple market morphism.

Let $c\in C$ and suppose that $c\in m_i'$ for some $i\in N\cup\{0\}$. Let $j>i$ and assume that $c\in m_i'\cap m_j'$. Since $c\in m_j'$ there is $m\in M$ with $c\in m$ such that $m\cap m_{j-1}'\neq\phi$. Also from $c\in m$ and $c\in m_i'$ we get that $m\subset m_{i+1}'$. Consequently, $i+1>j-1$ or $j=i+1$. Conversely, $m_i'\cap m_{i+1}'\neq\phi$ for each $i\in N\cup\{0\}$ since otherwise we would have a contradiction of the connectedness for M. ●

Theorem 6.7 can be considered as a fundamental structure theorem for market structures. It shows that for many purposes it is enough to consider canonical market structures. This way of reasoning presupposes that the assumptions used are independent of the market structure.

6.3. ECONOMIES OVER MARKET STRUCTURES

Now we are ready to introduce the general overlapping-markets economic models which provide the framework for our discussion of discrete time infinite horizon efficiency and optimality.

DEFINITION 6.8. Let m be a non-empty finite set of commodities. A <u>finite economy with commodities in m</u> is an array

$$\Psi^m=((X_i,P_i)_{i\in I^m},\ (Y_j)_{j\in J^m},\ \omega^m)$$

where

(i) I^m is a finite set of consumers, and for each $i \in I^m$,

$X_i \subset R^m$ is the consumption set of consumer i,

$P_i : X_i \to X_i$ is a preference relation on X_i

(ii) J^m is a finite set of producers, and for each $j \in J^m$,

$Y_j \subset R^m$ is the production set of producer j

(iii) $\omega^m \in R^m$ is the aggregate initial endowment in the economy

(iv) $I^m \cup J^m \neq \phi$. ●

An economy over a market structure is a triple(Ψ, C, M), where (C,M) is a market structure and Ψ is a map assigning to each $m \in M$ a finite economy Ψ^m with commodities in m.

For Ψ^m a finite economy, a consumption (production) bundle is an element $x_i \in X_i$ $(y_j \in Y_j)$. An allocation in Ψ^m is an array $z=((x_i)_{i \in I^m},(y_j)_{j \in J^m})_{m \in M}$ of consumption and production bundles. We assume that the reader is familiar with the concepts of Pareto optimal allocations and market equilibria for finite economies, otherwise the standard definitions will be special cases of those to follow. All of these concepts extend in a rather straightforward way to economies over market structures.

For $(\Psi,C,M,)$ an economy over a market structure and $i \in I^m$ a consumer in some market $m \in M$, a bundle x_i for i, that is, a vector in R^m, may be identified with a vector in R^C by assigning 0 to all coordinates not in m. This map from R^m to R^C is called the canonical embedding and the notation x_i will be used also for the image of x_i by the canonical embedding of R^m in R^C.

Let $I=\cup_{m \in M} I^m$, $J=\cup_{m \in M} J^m$, $H=I \cup J$ and let $m(i)$, $m(j)$ denote the smallest coordinate subspace $R^{m(i)}$ $(R^{m(j)})$ of R^C such that $R^{m(i)}$ $(R^{m(j)})$ contains X_i (Y_j). $m(h)$, $h \in H$, is called the carrier of h.

DEFINITION 6.9. Let (Ψ,C,M) be an economy over a market structure. An allocation in Ψ is a map z assigning to each $m \in M$ an allocation z^m in Ψ^m. The allocation z is feasible if

$x_i \epsilon X_i$, $i \epsilon I^m$, $y_j \epsilon Y_j$, $j \epsilon J^m$, each $m \epsilon M$

and

(1) $$\Sigma_{m\epsilon M}\Sigma_{i\epsilon I^m} x_{ic} = \Sigma_{m\epsilon M}\Sigma_{j\epsilon J^m} y_{jc} + \Sigma_{m\epsilon M} \omega^m_c$$

for all $c\epsilon C$. ●

A <u>price system</u> for (Ψ,C,M) is a vector $p\epsilon R^C$ with $p \geq 0$ for $c\epsilon C$. The price system is <u>essential</u> if for $h\epsilon H$ there is $c\epsilon m(h)$ such that $p_c > 0$, so that p induces a non-trivial price vector for each agent.

A <u>support</u> to a feasible allocation $\bar{z}$ for Ψ is an essential price system p such that

(i) for each $i\epsilon I$, $x_i \epsilon P(\bar{x}_i)$ implies $px_i \geq p\bar{x}_i$

(ii) for each $j\epsilon J$, $y_j \epsilon Y_j$ implies $p\bar{y}_j \geq py_j$

If it is also true that

(iii) for each $i\epsilon I$, $x_i \epsilon P(\bar{x}_i)$ implies $px_i > p\bar{x}_i$

then $(\bar{z},p)$ is a <u>market equilibrium</u> for Ψ.

The fundamental concept of this chapter and indeed of the whole book is that of Pareto optimality:

DEFINITION 6.10. Let (Ψ,C,M) be an economy over a market structure. An allocation z in Ψ is <u>Pareto optimal</u> if it is feasible and there is no feasible allocation z' such that for each $i\epsilon I^m$, $m\epsilon M$

(2) $$x_i' \epsilon P_i(x_i) \cup \{x_i\}$$

and $x_i' \neq x_i$ for some $i\epsilon I^m$ and some $m\epsilon M$. ●

In the definition of Pareto optimality we have assumed that each consumer must get either the original bundle or a bundle preferred to the original one if an allocation should be considered as a Pareto improvement over the original allocation. An alternative definition, replacing (2) above by

(3) $x_i' \epsilon \mathrm{cl}\{P_i(x_i) \cup \{x_i\}\}$ all i and $x_i' \epsilon P_i(x_i)$ some i

might seem more in accordance with the textbook definition of Pareto optimality in a finite economy. In this case, assuming that preferred sets are open, the two definitions are easily seen to be equivalent, whereas in our case the relationship is somewhat more subtle, we consider this question in the following chapter. At present we have opted for for the formulation (2) as the most intuitive in the general case with possibly incomplete preferences.

DEFINITION 6.11. Let (Ψ,C,M) be an economy over a market structure. An allocation z in Ψ is <u>weakly Pareto optimal</u> if it is feasible and there is no other feasible allocation $\bar{z}$ satisfying (2) such that

$$\{m\in M \mid \bar{z}^m \neq z^m\}$$

is finite. ●

We cannot extend the reasoning in finite economies to get a result on Pareto optimality of market equilibria - actually, the models of this book may be viewed as an outgrowth of the fact that market equilibria need not be Pareto optimal. What can be shown is a weak optimality property.

As usual, adopting the criterion of Pareto optimality for judging upon allocations in an economy implies that consumers preferences are taken as the ultimate measure of desirability. Since we want to treat special cases of economies involving only production we shall need two analogous concepts of production efficiency which may be applied to such economies.

DEFINITON 6.12. Let (C,M) be a market structure and $((Y_j)_{j\in J^m})_{m\in M}$ a family of production sets satisfying (ii) of Definition 6.8. A family $((\bar{y}_j)_{j\in J^m})_{m\in M}$ with $\bar{y}_j\in Y_j$ for $j\in J$ is <u>(weakly) production efficient</u> if there is no other family $((y_j)_{j\in J^m})_{m\in M}$ with $y_j\in Y_j$ (and $y_j=\bar{y}_j$ except for finitely many $j\in J$) such that

$$\Sigma_{m\in M}\Sigma_{j\in J^m}\, y_j > \Sigma_{m\in M}\Sigma_{j\in J^m}\, \bar{y}_j. \;●$$

It may be verified that if preferences of all consumers are strictly monotonic, so that a consumer in Ψ^m always prefers more of any of the commodities in m and each commodity c belongs to some $m(i)$, $i\in I$, then a Pareto optimal allocation is efficient. The proof is the same as for the corresponding result for finite economies.

We have already allowed for incomplete preferences and we choose to reduce the concept of (weak) production efficiency to (weak) Pareto optimality despite the fact that we then have to allow for preferences which are not open.

LEMMA 6.13. Let (C,M) be a market structure and $((Y_j)_{j\varepsilon J^m})_{m\varepsilon M}$ a family of production sets satisfying (ii) of Definition 6.8. A family $((\bar{y}_j)_{j\varepsilon J^m})_{m\varepsilon M}$ is (weakly) production efficient if and only if the allocation

$$((\bar{x}^m),(\bar{y}_j)_{j\varepsilon J^m})_{m\varepsilon M}$$

where

$$\bar{x}^m = \Sigma_{j\varepsilon J^m} \bar{y}_j + \omega^m$$

is (weakly) Pareto optimal in the economy with producers $((Y_j)_{j\varepsilon J^m})_{m\varepsilon M}$ and with a single consumer on each market mεM with consumption set R^m_+, preferences $P^m(x^m)=\{x^m\}+\dot{R}^m_+$ and initial endowment ω^m. ●

The proof of Lemma 6.13, which is just an application of the preceding definitions, is left to the reader. We will refer to an economy where there is only one consumer on each market described as in Lemma 6.13 as <u>a quasi production economy</u>.

THEOREM 6.14. Let (Ψ,C,M) be an economy over a market structure, and let (z,p) be a market equilibrium for Ψ. Then z is weakly Pareto optimal.

REMARK 6.15. If (Ψ,C,M) is a quasi production economy then the definition of market equilibrium ensures that $p_c>0$ for cεC.

<u>Proof</u>: Suppose that z is not weakly Pareto optimal , and let z'≠z be an allocation satisfying (2) and differing from z only on $m_1, \ldots, m_k \varepsilon M$.

Using (2) and the definition of a market equilibrium we get that

$$(4) \qquad \Sigma^k_{h=1} p^{m_h}(\Sigma_{i\varepsilon I^{m_h}} x'_i - \Sigma_{j\varepsilon J^{m_h}} y'_j) > \Sigma^k_{h=1} p^{m_h}(\Sigma_{i\varepsilon I^{m_h}} x_i - \Sigma_{j\varepsilon J^{m_h}} y_j)$$

for h=1,...,k

Now the vector in R^C defined by

$$\Sigma_{m\varepsilon M}(\Sigma_{i\varepsilon I^m} x'_i - \Sigma_{j\varepsilon J^m} y'_j) - \Sigma_{m\varepsilon M}(\Sigma_{i\varepsilon I^m} x_i - \Sigma_{j\varepsilon J^m} y_j)$$

has zero coordinates for $c \notin m_1 \cup \ldots \cup m_k$, so the expression

$$\Sigma_{m\varepsilon M}\, p^m[(\Sigma_{i\varepsilon I^m}\, x_i' - \Sigma_{j\varepsilon J^m}\, y_j') - (\Sigma_{i\varepsilon I^m}\, x_i - \Sigma_{j\varepsilon J^m}\, y_j)]$$

is well-defined, yielding by (4) a positive number. This means that

$$(\Sigma_{i\varepsilon I^m}\, x_{ic}' - \Sigma_{j\varepsilon J^m}\, y_{jc}') - (\Sigma_{i\varepsilon I^m}\, x_{ic} - \Sigma_{j\varepsilon J^m}\, y_{jc}) > 0$$

for some $c\varepsilon m_1 \cup \ldots \cup m_k$ with $p_c > 0$.

On the other hand, since $((x_i)_{i\varepsilon I^m}, (y_j)_{j\varepsilon J^m})_{m\varepsilon M}$ and $((x_i')_{i\varepsilon I^m}, (y_j')_{j\varepsilon J^m})_{m\varepsilon M}$ are both feasible we have

$$\Sigma_{m\varepsilon M}[(\Sigma_{i\varepsilon I^m}\, x_i' - \Sigma_{j\varepsilon J^m}\, y_j') - (\Sigma_{i\varepsilon I^m}\, x_i - \Sigma_{j\varepsilon J^m}\, y_j)] = 0$$

From this contradiction we conclude that z is Pareto optimal. ●

6.4. EXAMPLES

In this section we consider some particular cases of economies over market structures.

EXAMPLE 6.16. Finest Market Structure. Let (Ψ, C, M) be an economy over a market structure. Recall that for each $i\varepsilon I$ and $j\varepsilon J$ the sets m(i) and m(j) were defined so that $R^{m(i)}$ and $R^{m(j)}$ are the smallest coordinate subspaces of R^C containing X_i and Y_j respectively.

From the fact that (C,M) is a market structure it follows that

$$M' = \{m' \mid m' = m(i) \text{ or } m' = m(j) \text{ for some } i\varepsilon I, j\varepsilon J\}$$

is a market structure. This is the finest market structure such that the given collection of consumers and producers may be construed as an economy over a market structure. ●

EXAMPLE 6.17. General Production Models. Let $M = \{m_0, m_1, \ldots\}$ be a canonical market structure over a set C of commodities. A general production model is defined by a family $(Y^m)_{m\varepsilon M}$ of production sets $Y^m \subset R^m$

and initial endowments $\omega^m \epsilon R^m$. In the interpretation ω^m are the given supplies or demands of the commodities involved. Often it is assumed that $\omega^m=0$ except for $m=m_o$ and interest is focused upon what the production system can supply net to the consumers.

A program in the general production model is a family $(y^m)_{m\epsilon M}$ of production bundles $y^m \epsilon Y^m$ such that

$$\Sigma_{m\epsilon M}\omega^m+\Sigma_{m\epsilon M}y^m \geq 0$$

Here we have used the standard sign convention so that negative coordinates of y^m means that the corresponding commodity is used as input, positive coordinates means that the commodity is an output. The initial endowments may be incorporated in the production sets by defining new production sets $\bar{Y}^m=Y^m+\omega^m$.

(Weak) efficiency of the general production model may be reduced to the study of (weak) optimality by adding to each market a consumer, as in Lemma 6.13, but now with consumption set $X^m=R^m_+$.

If M has the particularly simple form of Example 6.4 , $\omega^m=0$ except for $m=m_o$, and each Y^m may be described by a production function, we are back to the production models of Part I. ●

EXAMPLE 6.18. Generation Models with Consumption and Production. A consumption model with several consumers on each market was introduced already in Part I for the market structure of Example 6.4. The general economy over a (canonical) market structure provides the obvious generalization of this model, allowing for many commodities at each date and for the presence of producers. ●

The models of the two previous examples are by far the most important cases of economies over market structures. It will be noticed that in both of them, the market structure involved was the canonical one. However, we have already encountered models where the market structure is not canonical (the double infinity production model in 2.4.), and below we consider some further examples. It must be conceded that such models tend to be somewhat artificial, and anyway the predominance of the canonical market structure was established by Theorem 6.7. and will be further stressed in the next section. The main function, then, of the abstract market structures is to state as clear as possible, without "irrelevant" details about time, infinite horizon etc. the properties common to the models which can be treated by the methods of this book.

EXAMPLE 6.19. Location. An example of an economy over a market structure not involving time is obtained by expanding on the classical theme of distinguishing goods by their location: Suppose that there is a countable number of different locations, in each of which a finite number of consumers and producers can trade in a market for a finite number of commodities. The locations are joined by the transportation technology, each of these transport technologies linking a finite number of locations.

We leave it to the reader to check that the situation sketched above can be formalized as an economy over a market structure, with markets corresponding either to locations or to transportation producers. ●

As an adequate formalization of a problem in regional economics, the above example may be somewhat deficient. Also there is little compelling reason to insist on a countable infinity of locations in the model. From the purely formal point of view, this model displays a peculiar feature of some interest: Although the market structure is connected, the markets containing consumers are mutually disjoint. We shall return to economies with this property in the following section.

6.5. MAPS AND MORPHISMS BETWEEN ECONOMIES OVER MARKET STRUCTURES

In Section 6.2 we saw that every market stucture can be transformed into a canonical market stucture of a particularly simple form. We will now show that this result can be extended to economies over market structures. Actually, a given map $\gamma:C\to C'$ gives rise to mappings between economies in two different ways; which we treat below.

I. Let $\eta:(C,M)\to(C',M')$ be a market morphism. η induces a map $f_\eta:R^{C'}\to R^C$ defined by

$$f_\eta(x)(c)=x(\eta(c))$$

for each $c\in C$, $x\in R^{C'}$. We denote by $f_\eta x f_\eta$ the map from $R^{C'}xR^{C'}$ to $R^C x R^C$ taking (x,y) to $(f_\eta(x),f_\eta(y))$. Also, whenever convenient we identify elements of R^m for $m\subset C$ with their images by the canonical embedding of R^m in R^C.

DEFINITION 6.20. Let (Ψ,C,M) and (Ψ',C',M') be economies over market structures. Let $I=\cup_{m\in M}I^m$, $J=\cup_{m\in M}J^m$ and define I', J' similiarly. We say that (Ψ,C,M) and (Ψ',C',M') are <u>comparable</u> if there is a market morphism

$$\eta:(C,M)\longrightarrow(C',M') \text{ and } \zeta:I'\longrightarrow I,\ \pi:J'\longrightarrow J$$

such that

(i) for $i'\in I'$, if $i=\zeta(i')$, then

$$X_i=f_\eta(X_{i'}),\ P_i=(f_\eta x f_\eta)(P_{i'}),\ \omega_i=f_\eta(\omega_{i'})$$

(ii) for $j'\in J'$, if $j=\pi(j')$ then

$$Y_j=f_\eta(Y_{j'}).$$

If η, ζ, π are all bijections, we say that (Ψ,C,M) and (Ψ',C',M') are <u>compatible</u>. ●

The comparability relation between economies over market structures is clearly reflexive and transitive but not in general symmetric. On the other hand, the compatability relation is an equivalence relation.

If (Ψ,C,M) and (Ψ',C',M') are compatible with maps (η,ζ,π), then the map f_η is bijective with $(f_\eta)^{-1}=f_\eta{}^{-1}$ and it induces a one-one correspondence between states in (Ψ,C,M) and states in (Ψ',C',M'). Also, η induces a map from prices $(p_{c'})_{c'\in C'}$ in (Ψ',C',M') to prices $(p_c)_{c\in C}$ in (Ψ,C,M) by

$$p_c=p_{\eta(c)}$$

for all $c\in C$. These maps take Pareto optimal allocations to Pareto optimal allocations and market equilibria to market equilibria.

THEOREM 6.21. Let (Ψ',C',M') be an economy over a market structure. Then there is a canonical market structure (C,M) and an economy Ψ over (C,M) such that

(i) (Ψ,C,M) and (Ψ',C',M') are compatible

(ii) there is a one-one correspondence between Pareto optimal allocation (Ψ,C,M) and (Ψ',C',M') as well as between market equilibria in (Ψ,C,M) and (Ψ',C',M'). ●

The proof of Theorem 6.21 is a straightforward application of the above definitions and is left to the reader.

Although the result of Theorem 6.21 may seem trivial it is quite important for the theory to follow since it allows us to concentrate on economies over canonical market structures: for many problems no loss of generality is involved in treating only these particular market structures rather than the general ones.

Actually, the ideas behind this simplification may be pushed somewhat further. In a canonical market structure, only two consecutive markets overlap, however, on each of these markets there may be several consumers and producers. Following a construction introduced by Clarke [1981] we may find an economy such that markets on which consumers are active never overlap, but which still has "the same" set of Pareto optimal states and market equilibria as the original one.

DEFINITION 6.22. Let (Ψ,C,M) and (Ψ',C',M') be comparable with maps η,ζ,π. We say that (Ψ,C,M) and (Ψ',C',M') are <u>quasi-compatible</u> if

(i) η is surjective, ζ is bijective

(ii) π is injective, and if $j\varepsilon J\setminus\pi(J')$, then $m(j)$ is the disjoint union of two sets m_1 and m_2 with $|m_1|=|m_2|=m$, and

$$Y_j=\{(y_1,-y_2)\,|\,y_1\varepsilon R^{m_1},y_2\varepsilon R^{m_2} \text{ and } y_1=-y_2\}. \;●$$

Thus if π is not surjective, then the remaining producers are trivial in the sense that they pass on a given vector from m_1 to m_2 - one may think of costless storing or transportation - but without actually transforming the commodities in a qualitative sense.

The following is a reformulation of Clarke's result:

THEOREM 6.23. Let (Ψ',C',M') be an economy over a market structure. Then there is a canonical market structure (C,M) and an economy Ψ over (C,M) such that

(i) (Ψ,C,M) and (Ψ',C',M') are quasi-compatible,

(ii) for $m_1, m_2 \in M$ if $I^{m_1} \neq \phi$, $I^{m_2} \neq \phi$, then $m_1 \cap m_2 = \phi$,

(iii) there is a one-one correspondence between Pareto optimal allocation (Ψ,C,M) and (Ψ',C',M') as well as between market equilibria in (Ψ,C,M) and (Ψ',C',M').

Proof: Without loss of generality we may assume that (C',M') is canonical, i.e.

$$M' = \{m_0', m_1', \ldots\}$$

with $m_i' \cap m_j' \neq \phi$ if and only if $i \in \{j-1, j, j+1\}$.

Let C be the set of pairs $(c',m') \in C' x M'$ such that $c' \in m'$, and let M consist of all sets

$$\{(c',m') \mid c' \in m'\}$$

for $m' \in M'$ together with the sets

$$\{(c',m_h) \mid h=i,i+1,\ c' \in m_i \cup m_{i+1}\}$$

for $i=0,1,\ldots$.

Finally, define $\gamma: C \rightarrow C'$ by $\gamma(c)=c'$ for $c=(c',m') \in C$. It is easily checked that (C,M) is a canonical market structure and that γ extends to a market morphism $\eta:(C,M) \rightarrow (C',M')$.

For each market in M of the form $\{(c',m') \mid c' \in m'\}$ let Ψ^m be the economy derived from $\Psi^{m'}$ in the obvious way (with the "same" commodities, same consumers and producers). For a market of the second type, let the economy Ψ^m consist of a single producer with production set

$$Y_j=\{(y_1,y_2)|(y_1,y_2)\varepsilon R^{m_i\cup m_{i+1}} \text{ and } y_1+y_2\leq 0 \}.$$

We leave it to the reader to check that the economy $\Psi=(\Psi^m)_{m\varepsilon M}$ over the market structure (C,M) is quasi-compatible with (Ψ',C',M').
Let $\bar{z}=(\bar{z}^m)_{m\varepsilon M}$ be a feasible allocation in Ψ'. Then for each i and $c'\varepsilon m_i\cup m_{i+1}$

$$\Sigma_{i\varepsilon I^{m_i}} \bar{x}_{ic'} -\Sigma_{j\varepsilon J^{m_i}} \bar{y}_{jc'}-\omega^m_{ic'} =$$
$$= \Sigma_{i\varepsilon I^{m_{i+1}}} \bar{x}_{ic'}-\Sigma_{j\varepsilon J^{m_{i+1}}} \bar{y}_{jc'}-\omega^m_{ic'}$$

Denote the vector with coordinates corresponding to the left hand side of this equality for $c'\varepsilon m_i\cup m_{i+1}$ by y_i, and assign the feasible production $(y_i,-y_i)$ to the producer on the market

$$\{(c',m_h)|h=i,i+1,\ c'\varepsilon m_i\cup m_{i+1}\}.$$

Together with the bundles obtained by applying f_η to bundles in Ψ' we obtain a feasible allocation in Ψ. Conversely, feasible allocations in Ψ give rise to feasible allocations in Ψ' in an obvious way. Now it is straightforward to check that Pareto optimal allocations in Ψ correspond to Pareto optimal allocations in Ψ', and that market equilibria in Ψ correspond to market equilibria in Ψ'. ●

We shall not make use of the result of Theorem 6.23 in the sequel.

II. Let (η,α) be a morphism. Recall that (η,α) induces the vectorspace homomorphism, $g_{\eta,\alpha}$, from $R^{(C)}$ to $R^{(C')}$ uniquely determined by its values on the vectors 1_c, $c\varepsilon C$, as these constitute a basis for $R^{(C)}$.

If (η,α) is a simple morphism then $g_{\eta,\alpha}$ is essentialy a change of the units of measurement; one old unit being equal to α_c new units. The morphisms of Chapter 4 were all of this type.

Any morphism, whether simple or not, induces a mapping between economies in the following way.

Let $(\Psi,C,M,)$ and (Ψ',C',M') be economies over market structures. Let $I=\cup_{m\varepsilon M}I^m$, $J=\cup_{m\varepsilon M}J^m$ and define I', J' similiarly. A mapping from (Ψ,C,M) to (Ψ',C',M') is then defined by a triple $((\eta,\alpha),\chi,\phi)$ where

$(\gamma,\iota):(C,M)\rightarrow(C',M')$, $\chi:I\rightarrow I'$, $\phi:J\rightarrow J'$, are such that

(i) (η,α) is a morphism

(ii) for $i\varepsilon I$,

$$X_{\chi(i)}=g_{\eta,\alpha}(X_i),\ P_{\chi(i)}=(g_{\eta,\alpha}xg_{\eta,\alpha})(P_i),\ \omega_{\chi(i)}=g_{\eta,\alpha}(\omega_i)$$

(iii) for $j\varepsilon J$,
$Y_{\phi(j)}=g_{\eta,\alpha}(Y_j)$.

Furthermore, allocations and supports in (Ψ,C,M) are mapped by $g_{\eta,\alpha}$ to allocations and supports in (Ψ',C',M') in an obvious way.

Many of the properties of economies, allocations and price systems are preserved by such a mapping $g_{\eta,\alpha}$. We will not make use of this in the sequel, rather we will apply similar ideas to reduced models in Chapter 7 and in Chapter 9.

CHAPTER 7: REDUCED MANY-GOODS MODELS

We have seen in Part I that the efficiency/optimality characterization in various one-good models can be given a uniform treatment by means of reduced models. The aim of the present chapter is to introduce reduced models in the many goods context as developed in the previous chapter.

Due to the abstract character of general economies over market structures, the reduced models will also be more complex than those of Part I. Moreover, care should be taken in the definition of the reduced model associated with a given economy and a given allocation: If economies (Ψ,C,M) and (Ψ',C',M') are compatible (in the sense of Definition 6.20) so that they are in a sense "the same", and if z and z' are allocations in the two economies which again correspond to each other, then the reduced model associated with (Ψ,C,M) and z should be "the same" as that associated with (Ψ',C',M') and z'. The need to verify these - fairly obvious as it might seem - properties of the reduced model will make it necessary to go through a certain amount of additional formalism.

The generalization of one-good models to models of economies over arbitrary market structures and in particular to models allowing for several goods at each date gives rise to complications of a rather technical nature. Therefore, although our reasoning in this chapter parallels that of Chapter 2, there are a number of additional details, some of which are of limited intrinsic interest. In spite of this, the main message to be delivered by this chapter is that the results of Chapter 2 carry over to the general case provided that we are willing to make a few, not very restrictive, additional assumptions.

7.1. REDUCED MODELS AND EFFICIENCY.

In the following, (C,M) will be an arbitrary market structure (cf. Definition 6.1).

DEFINITION 7.1. A reduced model over the market structure (C,M) is a family $\Sigma=(S^m)_{m\in M}$ of sets $S^m\subset R^m$, $m\in M$, satisfying

(i) S^m is closed, each $m\in M$,

(ii) $0\in S^m$, each $m\in M$,

(iii) $S^m+R^m_+\subset S^m$, each $m\in M$,

(iv) for any finite subset $\{m_1,\ldots m_k\}$ of M, the set $\Sigma_{i=1}^k S^{m_i}$ is closed. ●

The definition above should be compared with its counterpart in the one-good case (Definition 2.1). It is seen that the parts (i)-(iii) are immediate extensions to arbitrary dimensions of similar conditions stated for R^2. The condition (iv), however, is new (the sum of the sets S^{m_i} should be taken in the obvious way, i.e. after identifying $S^{m_i}\subset R^{m_i}$ with its image in R^C by the canonical embedding). The need for this type of property to be satisfied by a reduced model will become clear in the following.

Since (iv) implies (i), (i) is, strictly speaking, redundant. Condition (i) is retained only to facilitate comparison with Definition 2.1.

Before we proceed - essentially following the exposition of the theory in Section 2.1 - to define improvements and efficiency, we digress to consider a slight complication, which we shall face due to our desire to cover arbitrary market structures as well as applications to economies with consumers having incomplete preferences. The reader with no interest in this kind of problems may proceed directly to Definition 7.3.

DEFINITION 7.2. A generalized reduced model over a market structure (C,M) is a family $\Sigma_G=(S^m,S_*^m)_{m\epsilon M}$ of pairs (S^m,S_*^m) of subsets of R^m such that

(i) $(S^m)_{m\epsilon M}$ is a reduced model,

(ii) for each $m\epsilon M$, $S_*^m \subset S^m$. ●

The reason for introducing the sets S_*^m, as well as their interpretation, will become clear in the following. At this point we just want to sketch where the sets S_*^m enter.

Suppose that we construct a reduced model from an overlapping generations model with many goods at each date, working in the same way as we did in Section 2.3. In this situation, the sets S^m of the reduced model will be constructed as those deviations from the given allocation, which will make every member of the generation as least as well off, and S_*^m as the sets which make some member better off. The need to have both the sets S^m and S_*^m will arise only in the present general context. In the one-good overlapping generations model we were able to get around the problem in a different way.

DEFINITION 7.3. Let $\Sigma=(S^m)_{m\epsilon M}$ be a reduced model. A family $(\xi^m)_{m\epsilon M}$ is an improvement for Σ if

(i) $\Sigma_{m\epsilon M}\xi^m = 0$,

(ii) $\xi^m \epsilon S^m$ for all $m\epsilon M$, and $\xi^m \epsilon S^m+\dot{R}_+^m$ for some $m\epsilon M$. ●

The reduced model Σ is efficient if there is no improvement for Σ.

The summation in condition (i) of Definition 7.3 is performed in R^C. Note that due to (ii) of Definition 6.1, for every $c\epsilon C$, there is at most finitely many ξ^m such that $\xi_c^m \neq 0$.

At this point, the analogy between the concept of an improvement in the one-good and the many-good case is less perfect. Whereas in Chapter 2, the concept of an improvement had an almost immediate intuitive content, the present one is less clear - and actually not without certain drawbacks.

the point is that in applications we would like an improvement to mean that somebody is really better off, and for this the condition $\xi^m \epsilon S^m + \dot{R}^m_+$, its simplicity notwithstanding, may not be the right thing. This is why we will have to fall back occasionally on the generalized reduced models. The notion of improvement and efficiency may be defined here as well.

DEFINITION 7.4. Let $\Sigma_G = (S^m, S^m_*)_{m\epsilon M}$ be a generalized reduced model. A family $(\xi^m)_{m\epsilon M}$ is an <u>improvement</u> for Σ_G if

(i) $\Sigma_{m\epsilon M} \xi^m = 0$

(ii) $\xi^m \epsilon S^m$ for all $m\epsilon M$, and $\xi^m \epsilon S^m_*$ for some $m\epsilon M$.

The generalized reduced model Σ_G is <u>efficient</u> if there is no improvement for Σ_G. ●

In many cases, the sets S^m_* can be defined as $S^m + \dot{R}^m_+$, in which case it suffices to consider the ordinary reduced model $(S^m)_{m\epsilon M}$, since the notion of improvement for the generalized and the ordinary reduced models are then the same. There are important cases, however, where this identity does not obtain, and we return to these cases in due course.

Before we proceed, we give an example to clarify the relations between the notions of Part I and those defined in this section.

EXAMPLE 7.5. Let $\Sigma = (S_t)_{t\epsilon N}$ be a weakly efficient one-good reduced model (cf. Definitions 2.1 and 2.4), such that each S_t belongs to the family Ω (see Section 3.1). We can rewrite $(S_t)_{t\epsilon N}$ as a reduced model over the market structure

$$M = \{\{0,1\},\{1,2\},\ldots,\{t,t+1\},\ldots\} = \{\{t,t+1\} \mid t\epsilon N \cup \{0\}\}$$

considered in Example 6.4, as follows:

For $t=0$ we let $S^0 = R^2_+$; for $t>0$, $S^{m_t} = S_t$, where $m_t = \{t,t+1\}$. To see that $(S^{m_t})_{t\epsilon N\cup\{0\}}$ is actually a reduced model, over the market structure $(N\cup\{0\}, M)$, we need only verify condition (iv) of Definition 7.1.

Let $\{m_1,\ldots m_k\}$ be a finite subset of M. Clearly we may assume that $\{m_1,\ldots,m_k\} = \{\{t,t+1\},\{t+1,t+2\},\ldots,\{t+k,t+k+1\}\}$ for some t, and, by an induction argument, it suffices to consider the case of two overlapping markets. Suppose now that $(x^n)_{n\epsilon N}$, $(y^n)_{n\epsilon N}$ are sequences with $x^n\epsilon S^{m_t}$, $y^n\epsilon S^{m_{t+1}}$ such that x^n+y^n converges to some $z\epsilon R^3$. We will show that $z\epsilon S^{m_t}+S^{m_{t+1}}$

If (x^n) is unbounded, say $x_2^n\longrightarrow+\infty$, then $y_1^n\longrightarrow-\infty$, consequently the sequence $1/|y_1^n|(y_1^n,y_2^n)$, which by star-shapedness belongs to $S^{m_{t+1}}$, tends to $(-1,0)$ which is not in $S^{m_{t+1}}$, contradicting that $S^{m_{t+1}}$ is closed. If $x^n\longrightarrow-\infty$ we get a contradiction in a similar way. It follows that the sequences (x^n) and (y^n) are bounded, whence some subsequence converges to $x^o\epsilon S^{m_t}$, $y^o\epsilon S^{m_{t+1}}$, respectively, and we have $z=x^o+y^o\epsilon S^{m_t}+S^{m_{t+1}}$ (the reader will note the similarity of this argument with the proof of Lemma 3.1).

Now let $(\xi_t)_{t\epsilon N}$ be an improvement for $(S_t)_{t\epsilon N}$ (in the one-good case). Since the model is weakly efficient, there is $T\epsilon N$ such that $\xi_t=0$ for $t<T$ and $\xi_t>0$ for $t\geq T$. Therefore, the sequence $(\xi^t)_{t\epsilon N\cup\{0\}}$ defined by $\xi^o=(0,\xi_1)$, $\xi^t=(-\xi_t,\xi_{t+1})$ for $t\epsilon N$, satisfies $\xi^t\epsilon S^{m_t}$, all t, $\xi^t\epsilon S^{m_t}+\dot{R}_+^2$ for $t=T$. Thus $(\xi^t)_{t\epsilon N\cup\{0\}}$ is an improvement in the sense of Definition 7.3.●

Until now we have seen only trivial examples of reduced models not to speak of generalized reduced models. This situation will be remedied when in the following sections we discuss the general reduction procedure for economies over market structures.

We close this section by introducing some further notions, following Section 2.3 rather closely.

DEFINITION 7.6. Let $\Sigma=(S^m)_{m\epsilon M}$ $(\Sigma_G=(S^m,S_*^m)_{m\epsilon M})$ be a (generalized) reduced model. An improvement $(\xi^m)_{m\epsilon M}$ is <u>finite</u> if $\xi^m=0$ for all except finitely many $m\epsilon M$. The (generalized) reduced model (Σ_G) Σ is <u>weakly efficient</u> if there is no finite improvement for (Σ_G) Σ . ●

7.2*. INVARIANCE PROPERTIES OF REDUCED MODELS.

In the previous chapter we saw that an important role is played by morphisms between market structures, since this concept permitted us to go from a given, presumably complicated, market structure, to another and simpler one without changing the essentials of the problem.

Having introduced reduced models we are therefore led to investigate whether similar invariance properties can be established for them. We restrict our attention to simple morphisms since, firstly, these are the kind of transformations encountered in applications, and, secondly, invariance of efficiency will not hold for general morphisms.

Recall, that according to Definition 6.6, a simple morphism is a morphism $((\gamma,\iota),\alpha)$ with $\eta=(\gamma,\iota):(C,M)\to(C',M')$ such that $\gamma:C\to C'$ is a bijection.

Now we can define the transformation rule for reduced models under a simple morphism (η,α); let $h=g_{\eta,\alpha}$ be the linear map from $R^{(C)}$ to $R^{(C')}$ induced by (η,α), cf. Section 6.5. If γ is the identity map on C and $\alpha=1_C$ then h is the identity on $R^{(C)}$.

PROPOSITION 7.7. Let (C,M) and (C',M') be market structures, (η,α) a simple morphism. If $\Sigma=(S^m)_{m\varepsilon M}$ is a reduced model over (C,M), then the family $\Sigma(\eta,\alpha)=(S^{m'})_{m'\varepsilon M'}$, where

(1) $$S^{m'}=\Sigma_{m\varepsilon\iota^{-1}(m')}\ h(S^m)$$

if $m'\varepsilon\iota(M)$ and $S^{m'}=R_+^{m'}$ otherwise, is a reduced model over (C',M') . ●

The proof of Proposition 7.7 is a straightforward verification of the conditions (i)-(iv) in Definition 7.1. It should be noticed, however, that the result depends crucially on the fact that Σ satisfies (iv) of Definition 7.1.

The reduced model $\Sigma(\eta,\alpha)$ is called <u>the reduced model induced by</u> (η,α).

THEOREM 7.8. Let (η,α) be a simple morphism with $\gamma:C\to C'$, $\iota:M\to M'$ and $\Sigma(\eta,\alpha)$ the reduced model induced by (η,α). Then $\Sigma(\eta,\alpha)$ is (weakly) efficient if and only if Σ is (weakly) efficient.

<u>Proof</u>: Suppose that $(\xi^m)_{m\varepsilon M}$ is an improvement for Σ; define $(\bar{\xi}^{m'})_{m'\varepsilon M'}$ by

$$\bar{\xi}^{m'}=\Sigma_{m\varepsilon\iota^{-1}(m')}\ h(\xi^m) \quad \text{if } m'\varepsilon\iota(M) \text{ and } \bar{\xi}^{m'}=0 \text{ otherwise,}$$

then $(\bar{\xi}^{m'})_{m'\varepsilon M'}$ is easily seen to be an improvement for $\Sigma(\eta,\alpha)$. Thus $\Sigma(\eta,\alpha)$ efficient implies that Σ is efficient.

Conversely, suppose that $(\xi^{m'})_{m'\epsilon M}$ is an improvement for $\Sigma(\eta,\alpha)$. Then, for each $m'\epsilon\iota(M)$ we have $\xi^{m'}\epsilon S^{m'}$, that is,

$$\xi^{m'}=\Sigma_{m\epsilon\iota^{-1}(m')}\ h(\xi^{m})$$

for some ξ^{m}, $m\epsilon\iota^{-1}(m')$. Since $\{\iota^{-1}(m')|m'\epsilon M'\}$ is a partition of M, we get a family $(\xi^{m})_{m\epsilon M}$. It is easy to verify that $(\xi^{m})_{m\epsilon M}$ is an improvement for Σ. Thus Σ efficient implies $\Sigma(\eta,\alpha)$ efficient.

Clearly, if in the above arguments one of the improvements was finite, then so was the other one, showing that the result can be extended to cover weak efficiency as well. ●

The invariance of efficient reduced models under simple morphisms means that we may perform several operations of merging small markets into larger markets while still retaining all the important features of the reduced model. This implies that,for many purposes, we may without loss of generality restrict our attention to canonical market structures (cf. Example 6.5) which are more intuitive that general market structures.

The construction in Proposition 7.7 of a reduced model, using the map $g_{\eta,\alpha}$, induced by a simple morphism, and preserving efficiency, can be carried over to generalized reduced models. We give the general construction and leave it to the reader to check that a counterpart of Theorem 7.8 holds true with this construction.

Let (η,α) be a simple morphism.. For $\Sigma_G=(S^{m},S^{m}_{*})_{m\epsilon M}$ a generalized reduced model over (C,M), define $\Sigma_G(\eta,\alpha)=(\overline{S}^{m'},\overline{S}^{m'}_{*})_{m'\epsilon M'}$ by

$$(2)\qquad \begin{cases} \overline{S}^{m'}=\Sigma_{m\epsilon\iota^{-1}(m')}h(S^{m}) \text{ if } m'\epsilon\iota(M),\ S^{m'}=R^{m'}_{+} \text{ otherwise} \\ \text{and} \\ \overline{S}^{m'}_{*}=\cup_{\overline{m}\epsilon\iota^{-1}(m')}\left[h(S^{\overline{m}}_{*})+\Sigma_{m\epsilon\iota^{-1}(m')\setminus\{\overline{m}\}}\ h(S^{\overline{m}})\right] \end{cases}$$

if $m'\epsilon\iota(M)$, $\overline{S}^{m'}=\phi$ otherwise. Here we have used the convention that summing over the empty index set gives the zero vector.

7.3. THE REDUCED MODEL ASSOCIATED WITH AN ECONOMY AND AN ALLOCATION.

Having defined reduced models over arbitrary market structures, we may now proceed to consider the reduction procedure in the general case.

Let (Ψ,C,M) be an economy over a market structure (cf. Section 6.3), and let $\bar{z}=((\bar{x}_i)_{i\varepsilon I^m},(\bar{y}_j)_{j\varepsilon J^m})_{m\varepsilon M}$ be a feasible allocation in Ψ. For each $m\varepsilon M$, let

$$(3) \qquad S^m=\Sigma_{i\varepsilon I^m}\, cl(P_i(\bar{x}_i)-\bar{x}_i)-\Sigma_{j\varepsilon J^m}\,(Y_j-\bar{y}_j)$$

where the sum of sets over an empty index set is $\{0\}$ by definition.

Reasoning by analogy with Chapter 2, it would be tempting to define $(S^m)_{m\varepsilon M}$ right away as the reduced model associated with Ψ and z. However, there are two problems to be faced: (1) the collection $(S^m)_{m\varepsilon M}$ is not necessarily a reduced model in the sense of Definition 7.1, and (2) even if it is we still cannot be sure that Pareto optimality of z is equivalent to efficiency of $(S^m)_{m\varepsilon M}$, which after all is the main point in the construction of reduced models. We shall consider problem (1) at present and problem (2) in the following section.

First of all we introduce some standard assumptions on agents' characteristics in economies over market structures. Except for the notation and the context, these assumptions are well-known from the theory of ordinary finite horizon economies, whence we shall not discuss their interpretation.

Let (Ψ,C,M) be an economy over a market structure. If $h\varepsilon I\cup J$ is an agent, then the <u>carrier</u> of h, written m(h), is the smallest coordinate subspace of R^C containing the consumption set, if $h\varepsilon I$, or the production set, if $h\varepsilon J$, of agent h.

ASSUMPTION 7.9. The economy (Ψ,C,M) over the market structure (C,M) satisfies

(i) for each market m and consumer $i\varepsilon I^m$,

(a) X_i is a closed, convex subset of $R^{m(i)}$ with $X_i+R_+^{m(i)}\subset X_i$,

(b) for all $x_i \epsilon X_i$, $x_i \notin P_i(x_i)$, $P_i(x_i)$ is convex, and $P_i(x_i)+R_+^{m(i)} \subset P_i(x_i)$ (strict monotonicity),

(c) $\omega_i \epsilon X_i$

(ii) for each market m and producer $j \epsilon J^m$, Y_j is closed and convex, $0 \epsilon Y_j$ and $Y_j - R_+^{m(j)} \subset Y_j$.

In relation to (i.b) and (ii) above we note that monotonicity and free disposal can not reasonably be assumed for the whole commodity space which is R^C. Also since we want to allow for economies without consumers e.g quasi production economies, we have not made any openness assumption on preferences.Thus the assumptions above are rather weak, and as will be seen shortly, they are not sufficient to obtain the wanted correspondence between allocations in economies and reduced models. The problem is of course that we will have to show that property (i) - actually even the stronger version (iv) - of Definition 7.1 is fulfilled, a problem which in its turn reduces to showing closedness of a sum of closed sets.

We shall need the concept of the asymptotic cone $A[B]$ of a subset B of R^m, where m is a finite subset of C. For its definition, as well as for some properties of asymptotic cones, the reader is referred to Debreu [1959], p.22-23. For the sake of completeness we note that if m' is a finite subset of C and $m \subset m'$, then $A[B]$ taken in $R^{m'}$ is the same set as - strictly speaking, the image by the canonical embedding of - $A[B]$ taken in R^m. Thus we need not be explicit in the following on the exact finite-dimensional subspace of R^C in which the operation of taking the asymptotic cone is performed.

Let $\{C_i | i \epsilon I\}$ be an indexed family of cones in R^C such that each C_i is contained in some finite-dimensional coordinate subspace of R^C. The family $\{C_i | i \epsilon I\}$ is <u>positively semi-independent</u> if for any finite subset $\{i_1, \dots i_r\}$ of I, if $x_{i_j} \epsilon C_{i_j}$, $j=1,\dots,r$, and $\Sigma_{j=1}^r x_{i_j}=0$, then $x_{i_j}=0$ for every j. Without proof we state the following lemma which is a reformulation of 1.9(9) in Debreu [1959]

LEMMA 7.10. Let $\{B_i | i \epsilon I\}$ be an indexed family of closed subsets of R^C such that each B_i is contained in finite-dimensional coordinate subspace of R^C. If $\{A[B_i] | i \epsilon I\}$ is positively semi-independent, then $\Sigma_{j=1}^r B_{i_j}$ is closed for any finite subset $\{i_1, \dots, i_r\}$ of I. ●

Now the following result is quite straightforward:

THEOREM 7.11. Let (Ψ,C,M) be an economy over a market structure satisfying Assumption 7.9, and let $\bar{z}=((\bar{x}_i)_{i\varepsilon I^m},(\bar{y}_j)_{j\varepsilon J^m})_{m\varepsilon M}$ be a feasible allocation in Ψ. If the family

$$\{A[P_i(\bar{x}_i)]\,|\,i\varepsilon I\}\cup\{A[Y_j]\,|\,j\varepsilon J\}$$

is positively semi-independent, then the family $(S^m)_{m\varepsilon M}$ given by (3) is a reduced model.

Proof: This is a simple check of conditions (i)-(iv) in Definition 7.1. Condition (ii) follows from the fact that $\bar{z}$ is feasible; (iii) is a consequence of Assumption 7.9,(i).b and (ii). Finally, (i) follows from (iv) which again follows from Lemma 7.10. ●

Conditions on an economy stated in terms of asymptotic cones tend to be rather difficult to interpret. In order to show that the conditions of Theorem 7.11 are not extremely restrictive we consider a special case which nevertheless is sufficiently general to cover the models of Part I.

Let (Ψ,C,M) be an economy over a market structure which we assume to be canonical. For ease of notation we assume that each m in M can be partitioned in sets m_{tF},m_{tL} with $m_{tL}=m_t\cap m_{t+1}$, $t=0,1,2,...$ and that $m_o=m_{oL}$. Assume furthermore that consumption sets are bounded from below, and that for each producer $j\varepsilon J^m,m\varepsilon M$, the production set Y_j is a subset of $-R^{m_F}xR^{m_L}$ such that for all $y_j\varepsilon Y_j$, if $y^{m_L}\not\in -R^{m_L}_+$ then $y^{m_F}\neq 0$. The latter assumption has a straightforward interpretation when markets are thought of as covering two consecutive time periods (dates), the First and the Last, namely that no output can be obtained in the first period. This is possible only in the second period following some input made in the first. This kind of assumption on the technology was implicit in all of Part I.

Returning to the semi-independence condition, assume that for each m, the set S^m is already known to be closed (so that it remains only to verify condition (iv)). If $\{t_1,...,t_r\}$ is a finite subset of M, $x_{t_i}\varepsilon A[S_{t_i}],x_{t_i}\neq 0$, $i=1,...,r$, and $\Sigma_{i=1}^r x_{t_i}=0$, then for t_1 the smallest of the indices $t_1,...,t_r$ we must have either $t_1=0$, in which case $x_{t_1}\varepsilon R^{m_{t_1}}$ and $x_{t_2}\varepsilon R^{m_{t_2}}$, contradicting that their sum is 0, or $t_1>0$ in which case $x_{t_1}=(0,x_{t_1})\varepsilon\{0\}xR^{m_{t_1}}$, and the argument runs as before. Thus, we obtain condition (iv) as a result of a particular set of rather reasonable assumptions on the technology. Other assumptions might be investigated as well; we shall not pursue this topic any further.

7.4. PARETO OPTIMALITY AND EFFICIENCY OF THE REDUCED MODEL.

As pointed out already, the reason for our construction of reduced models is that they allow us to analyze the efficiency/optimality problem in a uniform and transparent way. Therefore, it is important that Pareto optimal allocations translate to efficient reduced models, and vice versa. This connection between optimality properties of alternative representations of the model was almost trivial in the one-good case. However, as in the previous section, we encounter some difficulties, mainly of notational character, when going from one-good to several goods models.

Consider an economy (Ψ,C,M) over a market structure and a feasible allocation $\bar{z}=((\bar{x}_i)_{i\in I^m},(\bar{y}_j)_{j\in J^m})_{m\in M}$ in Ψ. If $\bar{z}$ is not Pareto optimal, then there is a feasible allocation $z=((x_i)_{i\in I^m},(y_j)_{j\in J^m})_{m\in M}$ such that

$$(4)\qquad \begin{cases} x_i\in P_i(\bar{x}_i)\cup\{\bar{x}_i\}, \text{ all } i\in I \\ \\ x_{i'}\in P_{i'}(\bar{x}_{i'}) \quad \text{some } i'\in I. \end{cases}$$

Using (2) it is clear that $(\xi^m)_{m\in M}$ defined by

$$(5)\qquad \xi^m=\Sigma_{i\in I^m}\,(x_i-\bar{x}_i)-\Sigma_{j\in J^m}\,(y_j-\bar{y}_j)$$

satisfies most of the conditions imposed on an improvement for the reduced model $(S^m)_{m\in M}$ associated with (Ψ,C,M) and $\bar{z}$. Indeed, the condition $\Sigma_{m\in M}\xi^m=0$ follows from the fact that both z and $\bar{z}$ are feasible; that $\xi^m\in S^m$ for each $m\in M$ is obvious from (2), so it remains only to verify that $\xi^{m'}\in S^{m'}+\dot{R}^{m'}_+$ for some $m'\in M$.

This, in its turn, does not follow from the construction alone, but it will emerge as an easy consequence once we put the standard assumptions on the original economy.

PROPOSITION 7.12. Let (Ψ,C,M) be an economy over a market structure satisfying Assumption 7.9. Let $\bar{z}=((\bar{x}_i)_{i\varepsilon I^m},(\bar{y}_j)_{j\varepsilon J^m})_{m\varepsilon M}$ be an allocation in Ψ and $(S^m)_{m\varepsilon M}$ the reduced model asssociated with Ψ and $\bar{z}$. Then $\bar{z}$ is Pareto optimal if $(S^m)_{m\varepsilon M}$ is efficient.

Proof: If $x_i'\varepsilon P_i(\bar{x}_i)$, then $x_i'=x_i''+u$ for some $x_i''\varepsilon P_i(\bar{x}_i)$, $u\varepsilon \dot{R}^{m(i)}$, consequently $x_i'\varepsilon P_i(\bar{x}_i)+\dot{R}^m_+$. This together with the above remarks proves the proposition. ●

The converse of Proposition 7.12 does not follow without additional assumptions. The problem arising in the general case is the following: Suppose that $(S^m)_{m\varepsilon M}$ is the reduced model associated with a feasible allocation $\bar{z}$ in Ψ. If $(S^m)_{m\varepsilon M}$ is not efficient, then we can find an improvement $(\xi^m)_{m\varepsilon M}$ for $(S^m)_{m\varepsilon M}$, and using (3) we can find a feasible allocation $z=((x_i)_{i\varepsilon I^m},(y_j)_{j\varepsilon J^m})_{m\varepsilon M}$ such that

$$(6)\qquad \begin{cases} x_i\varepsilon \operatorname{cl}P_i(\bar{x}_i), & \text{all } i\varepsilon I \\ x_{i'}\varepsilon P_{i'}(\bar{x}_{i'}), & \text{some } i'\varepsilon I. \end{cases}$$

This, however, is in general strictly weaker than (4) and we cannot conclude that z is not Pareto optimal.

In most of the situations to be encountered in applications, the conditions (4) can be derived from (6) by simple arguments. We consider some particular cases.

EXAMPLE 7.13. (Ψ,C,M) is a quasi-production economy (cf. Section 6.4), so that for each market m there is exactly one consumer with $P^m(x^m)=\{x^m\}+\dot{R}^m_+$ for every $x^m\varepsilon R^m$. For $\bar{z}$ a feasible allocation, the generalized reduced model as defined by (3) and (7) has the form $(S^m,S^m+\dot{R}^m_+)_{m\varepsilon M}$. Clearly, the improvements for $(S^m)_{m\varepsilon M}$ coincides with the improvements for $(S^m,S^m+\dot{R}^m_+)_{m\varepsilon M}$. ●

EXAMPLE 7.14. Let (Ψ,C,M) be an economy over a market structure, satisfying Assumption 7.9 and such that each good is consumed by some consumer. Suppose, futhermore, that for any consumer $i\varepsilon I^m$, the preferences can be represented by a continuous utility function, i.e. there is $u_i:X_i\to R$ such that $x_i\varepsilon P_i(x_i')$ if and only if $u_i(x_i)>u_i(x_i')$. In this case, where preferences are complete, it is customary to accept a feasible allocation z as a Pareto improvement over $\bar{z}$ if

$$(7)\qquad \begin{cases} u_i(x_i) \geq u_i(\bar{x}_i), & \text{all } i \in I \\ u_i(x_i) > u_i(\bar{x}_i), & \text{some } i \in I, \end{cases}$$

accepting indifference for some consumers. Condition (7) is obviously weaker than (4). However, this alternative notion of improving may be built into our formalism by redefining preferred sets at $\bar{x}_i$ to be

$$P''_i(\bar{x}_i) = \text{cl}\, P_i(\bar{x}_i) \setminus \{\bar{x}_i\}$$

for each $i \in I$. In particular we note that P''_i is an irreflexive preference relation. Let $(S^m)_{m \in M}$ be the reduced model associated with $\bar{z}$. Defining the generalized reduced model using these "new" preferences, we see that it has the form

$$(S^m, S^m \setminus \{0\})_{m \in M}$$

An improvement for $(S^m)_{m \in M}$ induces an allocation z' satisfying (4). By construction of the preferences P''_i this implies the existence of an allocation z satisfying (7). ●

For the following result, we assume that the market structure is canonical. As we have seen, this does not in itself imply a loss of generality, since any market structure may be "made" canonical (by a simple market morphism), but on the other hand the assumptions to be made on the economy together with the assumptions on the market structure may impose a restriction.

PROPOSITION 7.15. Let (Ψ, C, M) be an economy over a canonical market structure. Suppose that Ψ satisfies Assumption 7.9 and that for each $m \in M$, $I^m \neq \phi$, $m(i) = m$ and $P_i(x_i)$ is open in $R^{m(i)}$ for $i \in I^m, x_i \in X_i$. If $(S^m)_{m \in M}$ is the reduced model associated with the feasible allocation $\bar{z} = ((\bar{x}_i)_{i \in I^m}, (\bar{y}_j)_{j \in J^m})_{m \in M}$, then $(S^m)_{m \in M}$ is efficient if $\bar{z}$ is Pareto optimal.

The proof of Proposition 7.15 is straightforward, but tedious: Define a feasible allocation by splitting the gain over all the (countably many) consumers. To avoid complicated notation we give the proof only for the case where there is exactly one consumer on each market.

Proof of Proposition 7.15: Suppose that $(S^m)_{m\epsilon M}$ is not efficient, and let

$$z=((x_i)_{i\epsilon I^m},(y_j)_{j\epsilon J^m})_{m\epsilon M}$$

be a feasible allocation satisfying (6). We show that z may be transformed to a feasible allocation

$$z'=((x_i')_{i\epsilon I^m},(y_j')_{j\epsilon J^m})_{m\epsilon M}$$

with $y_j=y_j'$, all $j\epsilon J$, satisfying (4).

Since (C,M) is canonical we may write $M=\{m_o,m_1,.....\}$. Let τ be an index such that for $i\epsilon I^{m\tau}$, $x_i\epsilon P_i(\bar{x}_i)$. Choose $u\epsilon R^m_{++}$ such that $x_i-u\epsilon P_i(\bar{x}_i)$, and write $u=(u^F,u^o,u^L)$ with $u^F\epsilon R^{m_F}_{++}$, $u^o\epsilon R^{m\setminus(m_F\cup m_L)}_{++}$, $u^L\epsilon R^{m_L}_{++}$, where $m_F=m_t\cap m_{t-1}$, $m_L=m_t\cap m_{t+1}$. Let $x_i''=x_i-u+u^o$ for $i\epsilon I^{m\tau}$. For $i\epsilon I^{m\tau-1}$, let $x_i''=x_i+u^F$, and for $i\epsilon I^{m\tau+1}$, let $x_i''=x_i+u^L$. Now we have $x_i''\epsilon P_i(x_i)$ for all i in $I^{m\tau-1}$, $I^{m\tau}$, and $I^{m\tau+1}$, and we may repeat the argument for $\tau-1$ and $\tau+1$. Proceeding in this way (the part of the argument going backwards from τ to $\tau-1$ terminating in a finite number of steps), we define inductively a sequence $(x_i'')_{i\epsilon I^{m_t}}$ such that (4) is fulfilled. ●

The idea of transferring goods between generations, which is essentially what makes the proof of Proposition 7.15 work, may be applied also in situations more general than this, but it does not apply universally. The reader may construct examples of convex sets $S^m\subset R^{m_1}xR^{m_2}$ such that for some $x\epsilon bdS^m$, $u\epsilon \dot{R}^{m_1}_+$ we have $x+(u,0)\epsilon bdS^m$, that is, there is a lack of substitutability between goods in m_1 and goods in m_2.

To cover the general case with lack of substitution possibilities we have already introduced the concept of a generalized reduced model; it remains only to define the generalized reduced model associated with an economy (Ψ,C,M) and an allocation $\bar{z}=((\bar{x}_i)_{i\epsilon I^m},(\bar{y}_j)_{j\epsilon J^m})_{m\epsilon M}$ as $(S^m,S^m_*)_{m\epsilon M}$, where S^m is given by (3), each $m\epsilon M$, and

$$(8)\qquad S^m_*=\cup_{i\epsilon I^m}\ [(P_i(\bar{x}_i)-\bar{x}_i)+\Sigma_{i'\epsilon I^m\setminus\{i\}}(P_{i'}(\bar{x}_i)\cup\{\bar{x}_{i'}\})]-\Sigma_{j\epsilon J^m}\ (Y_j-\bar{y}_j)$$

We note that $(S^m,S^m_*)_{m\epsilon M}$ is a generalized reduced model under the assumptions considered in Theorem 7.11.

We leave it to the reader to verify the follwing proposition.

PROPOSITION 7.16. Let (Ψ,C,M) be an economy over a market structure, z an allocation, and let $(S^m,S^m_*)_{m\varepsilon M}$ be the generalized reduced model associated with Ψ and $\bar{z}$. Then $\bar{z}$ is Pareto optimal if and only if $(S^m,S^m_*)_{m\varepsilon M}$ is efficient.●

In finite horizon models, the two alternative notions of a Pareto improvement can be shown to be equivalent when preferences satisfy some (weak) continuity assumptions. This does not carry over immediately; it is crucial - as can be seen in the proof of Proposition 7.15 - that there are consumers on all markets, or otherwise that the production possibilities exhibit sufficiently strong, intertemporal, substitution properties, allowing us to get more of all future goods as output whenever more of all goods today are offered as input.

7.5.* THE EXISTENCE OF SUPPORTS.

For one-good models supporting prices were defined inductively. We demanded in Section 2.1 that prices should be positive. In relation to efficiency considerations there was no loss in generality in doing so, since if a reduced model Σ is supported by $(p_t)_{t\varepsilon N}$ with $p_t=0$ infinitely often, then Theorem 3.9 immediately gives us that the model is efficient.

For many-goods models the situation is more complicated. Let (Ψ,C,M) be an economy over a market structure and consider a weakly efficient allocation z. One may think of several different notions of a support at z. If, for example, we consider a family $(p_c)_{c\varepsilon C}$ satisfying $p_c\geq 0$ for $c\varepsilon C$ and $p_c>0$ for some $c\varepsilon C$, then this may yield a non-trivial support only to very few of the individual actions involved. For our purposes it would be natural to demand at least, that for all $m\varepsilon M$ there is $c\varepsilon m$ such that $p_c\neq 0$, that is, p^m is non-zero for every market m. This requirement, however, is not independent of the market structure considered.In view of this we introduced the notion of an essential price system in Section 6.3, demanding that for each agent $h\varepsilon H$, $p_c\neq 0$ for some $c\varepsilon m(h)$. This condition is equivalent to the one discussed before when M is the finest market structure compatible with (Ψ,C,M).

Supports for (generalized) reduced models are defined in the obvious way:

DEFINITION 7.17.

(a) Let $\Sigma=(S^m)_{m\in M}$ be a reduced model over a market structure (C,M). A family $(p_c)_{c\in C}$ is called a <u>support</u> for Σ if

(i) for each $m\in M$, $p_c>0$ for some $c\in m$

(ii) for each $m\in M$, $p^m=(p_c)_{c\in m}$ satisfies $p^m x\geq 0$, all $x\in S^m$.

(b) Let $\Sigma_G=(S^m,S^m_*)_{m\in M}$ be a generalized reduced model over (C,M). A family $(p_c)_{c\in C}$ is a <u>support</u> for Σ_G if it is a support for $(S^m)_{m\in M}$. ●

In proving the existence of a support to a Pareto optimal allocation in a finite economy the procedure is to obtain at first a non-null price system such that consumer actions are expenditure minimizing relative to the price system. One obtains then a quasi-equilibrium (cf. Debreu [1962]) and using some further assumptions, one proves that actions are preference maximizing. For producers profit maximization is proved directly.

Proofs of existence of an equilibrium in a finite economy follow a similar route. First, the existence of a quasi-equilibrium is shown. Then using assumptions like "indirect resource-relatedness", "strictly positive endowments", etc., the quasi-equilibrium is shown to be an equilibrium. Cf. Arrow and Hahn [1971], Bergstrom [1976], Debreu [1962].

For economies over market structures almost the same techniques as in finite economies can be used for proving the existence of a support to a weakly Pareto optimal allocation or the existence of an equilibrium. By considering larger and larger finite subeconomies, a support or an equilibrium is found as a limit. We shall not treat in detail the problem of existence of equilibrium here. The reader is referred to Balasko, Cass and Shell [1980], Borglin and Keiding [1981] and Wilson [1981].

The suggested way of showing the existence of a support for an allocation in an economy over a market structure can be carried out only under some additional assumptions. The reason for this is that the price systems obtained in finite subeconomies might converge to 0. Hence it is necessary to introduce assumptions which ensure, for each finite subeconomy, that the price system is "sufficiently non-null", so that in the limit we get a price system which is a support for the actions of the agents.

Connectedness of the market structure does not preclude that some markets contain only goods which are of no use to the rest of the economy. As a consequence, the economy may split into several, essentially unrelated, subeconomies. To avoid this, we shall use a weak version of Malinvaud's non-tightness condition originally formulated for economies involving only production, cf. Malinvaud [1953].

DEFINITION 7.18. Let (Ψ,C,M) be an economy over a market structure with M the finest market structure compatible with Ψ. Furthermore, let

$$\bar{z}=((\bar{x}_i)_{i\varepsilon I^m},(\bar{y}_j)_{j\varepsilon J^m})_{m\varepsilon M}$$

be a feasible allocation, and $\bar{M}$ a finite subset of $\bar{M}$. The allocation $\bar{z}$ is <u>weakly non-tight</u> in $\bar{M}$ if

(a) for all $m',\bar{m}\varepsilon\bar{M}$ there exist $u^{\bar{m}}\varepsilon R^{\bar{m}}_{+}$, $h'\varepsilon I^{m'}\cup J^{m'}$, $\bar{h}\varepsilon I^{\bar{m}}\cup J^{\bar{m}}$, and an allocation $z=((x_i)_{i\varepsilon I^m},(y_j)_{j\varepsilon J^m})_{m\varepsilon M}$ such that

(i) $x_i\varepsilon clP_i(\bar{x}_i)$ for $i\varepsilon I\cup J$ and $y_j\geq\bar{y}_j$ for $h\varepsilon(I\cup J)\setminus\{\bar{h}\}$

(ii) $x_{h'}\varepsilon intP_{h'}(\bar{x}_{h'})$ if $h'\varepsilon I^{m'}$, $y_{h'}\varepsilon\{\bar{y}_{h'}\}+R^{m(h')}_{++}$ if $h'\varepsilon J^{m'}$

(iii) $x_h=\bar{x}_h$ and $y_h=\bar{y}_h$ for h such that $m(h)\notin\bar{M}$

(iv) $\Sigma_{m\varepsilon M}\Sigma_{i\varepsilon I^m}x_i-\Sigma_{m\varepsilon M}\Sigma_{j\varepsilon J^m}y_j=\Sigma_{m\varepsilon M}\Sigma_{i\varepsilon J^m}\omega_i+u^{\bar{m}}$.

The allocation $\bar{z}$ is <u>weakly non-tight</u> if there is an increasing sequence of finite subsets M_k of M such that $\cup_{k\varepsilon N}M_k=C$ and $\bar{z}$ is weakly non-tight in M_k for each $k\varepsilon N$. ●

The non-tightness assumption can be interpreted as stating that additional goods available on some market may be transferred to goods on another market by a chain of either productions or exchanges, meaning that substitution is possible for both consumers and producers. Thus non-tightness is one of several notions of the markets being related in an economic sense. We have encountered similar notions in the discussion leading to Proposition 7.12. Still another one will be introduced in Definition 7.21.

For some problems, for example the discussion of measures of curvature, it is essential that all prices be positive. Weak non-tightness does not ensure this. Hence we will also make use of the following stronger concept of non-tightness:

DEFINITION 7.19. The allocation $\bar{z}=((\bar{x}_i)_{i\epsilon I^m},(\bar{y}_j)_{j\epsilon J^m})_{m\epsilon M}$ is <u>strongly non-tight</u> if the conditions of Definition 7.18 are met with

"(a') for all $m',m\epsilon M$ and all $u^{\bar{m}}\epsilon \dot{R}_+^{\bar{m}}$ ".

substituted in the first line of (a) and "strongly non-tight" substituted for "weakly non-tight". ●

Strong non-tightness means that every good on market $\bar{m}$ is valuable, directly or indirectly, to producers or consumers in other markets and weak non-tightness that some combination of goods from market $\bar{m}$ has this property. Now we can state the following result:

THEOREM 7.20. Let (Ψ,C,M) be an economy over a market structure, where M is the finest market structure compatible with Ψ. Suppose that (Ψ,C,M) satisfies Assumption 7.9, and let $\bar{z}=((\bar{x}_i)_{i\epsilon I^m},(\bar{y}_j)_{j\epsilon J^m})_{m\epsilon M}$ be a weakly Pareto optimal allocation. If $\bar{z}$ is weakly non-tight, then there is an essential support $(p_c)_{c\epsilon C}$ to $\bar{z}$. If, furthermore, $\bar{z}$ is strongly non-tight, then $p_c>0$ for each $c\epsilon C$.

<u>Proof</u>: Since $\bar{z}$ is weakly non-tight there is an increasing sequence of sets $(M_k)_{k\epsilon N}$ such that $\bar{z}$ is weakly non-tight in M_k. for each k. Let $h''\epsilon I\cup J$ be such that $m(h'')\epsilon M_1$. Choose $k\epsilon N$, let $M_k=\bar{M}$ and

$$I(\bar{M})=\{i\epsilon I \mid m(i)\epsilon\bar{M}\}$$

$$J(\bar{M})=\{j\epsilon J \mid m(j)\epsilon\bar{M}\}$$

$$C(\bar{M})=\{c\epsilon C \mid c\epsilon m \text{ for some } m\epsilon\bar{M}\}$$

Define

$$Z(\overline{M})=\Sigma_{i\varepsilon I(\overline{M})}\, cl(P_i(\overline{x}_i)-\overline{x}_i) - \Sigma_{j\varepsilon J(\overline{M})}\,(Y_j-\overline{y}_j)$$

It is easy to check that $Z(\overline{M})$ is convex. From (i) and (ii) of Assumption 7.9 follows, since $\overline{z}$ is weakly Pareto optimal that

$$Z(\overline{M})\cap(-R_{++}^{C(\overline{M})})=\phi$$

Thus the sets $Z(\overline{M})$ and $-R_{++}^{C(\overline{M})}$ may be separated, that is there is $p\varepsilon R^C$, $p_c\neq 0$ for some $c\varepsilon C(\overline{M})$ such that

$$pz\geq 0 \text{ for } z\varepsilon Z(\overline{M})$$

$$p_c\geq 0 \text{ for } c\varepsilon C(\overline{M})$$

Let $m'\varepsilon\overline{M}$ be such that $p_c>0$ for some $c\varepsilon m'$. We will now show that for each $m\varepsilon\overline{M}$ there is $c\varepsilon m$ such that $p_c>0$.

Since $\overline{z}$ is weakly non-tight in $\overline{M}$ we may for any $\overline{m}\varepsilon\overline{M}$ find $u^{\overline{m}}\varepsilon R^{\overline{m}}$ and an allocation $z=((x_i)_{i\varepsilon I^m},(y_j)_{j\varepsilon J^m})_{m\varepsilon M}$ satisfying (i)-(iv) of Definition 7.18. Then

$$pu^{\overline{m}}=p\,[\Sigma_{i\varepsilon I}(x_i-\overline{x}_i) - \Sigma_{j\varepsilon J}(y_j-\overline{y}_j)]$$

by (iv) of Definition 7.18. (iii) of the same definition ensures that the sums occuring are finite.

By (i) and (ii) of Definition 7.18

$$p\Sigma_{i\varepsilon I}(x_i-\overline{x}_i)>0$$

$$p\Sigma_{j\varepsilon J}(y_j-\overline{y}_j)<0$$

with at least one strict inequality. It follows that $pu^{\overline{m}}>0$, which shows that p is non-null on $\overline{m}$. Since $\overline{m}$ and $\overline{M}$ were arbitrary it follows that p is non-null on any market $m\varepsilon M$.

In passing we note that if $\overline{z}$ is strongly non-tight, an analogous argument shows that $p_c>0$ for $c\varepsilon C(\overline{M})$.

Let $P(\overline{M})$ be the set of all $p\varepsilon R^C$ such that

$pz\geq 0$ for $z\varepsilon Z(\overline{M})$ and

$\|p^{m'}\|=1$ for some fixed $m'\varepsilon\overline{M}$

We will now show that there are constants $(K_c)_{c\varepsilon C(\overline{M})}$ such that $p_c\leq K_c$ for $c\varepsilon C(\overline{M})$, $p\varepsilon P(\overline{M})$.

To this end assume that there are price systems $p^n\varepsilon P(\overline{M})$, $n\varepsilon N$, such that p^n_c tends to infinity for some $c\varepsilon C(\overline{M})$. Consider the price systems $q^n=p^n/\|p^n\|$, $n\varepsilon N$. Without loss of generality we may assume that q^n converges to $q\varepsilon R^{C(\overline{M})}$, $q\neq 0$. Each q^n supports $Z(\overline{M})$ and since the supports to $Z(\overline{M})$ form a closed set; q supports $Z(\overline{M})$. But $\|q^{m'}\|=0$ and this contradicts the property that $q_c>0$ for some $c\varepsilon m'$. Hence $p\varepsilon Z(\overline{M})$ implies $p_c\leq K_c$ for some $K_c\varepsilon R$ for $c\varepsilon C(\overline{M})$.

We have now shown that for a given $k\varepsilon N$, $P(M_k)$ is non-empty and that there are constants K_c, $c\varepsilon m$ and $m\varepsilon M_k$, such that $p\varepsilon P(M_k)$ implies $p_c<K_c$ for $c\varepsilon m$ and $m\varepsilon M_k$.

It follows that

$$\cap_{k=1}^{\infty} P(M_k) = \cap_{k=1}^{\infty} \Pi(M_k)$$

where

$$\Pi(M_k) = P(M_k)\cap\{p\varepsilon R^C \mid p_c\leq K_c \text{ for } c\varepsilon C\}$$

$\Pi(M_k)$, $k\varepsilon N$, is non-empty and compact in the product topology. Since $\Pi(M_{k+1})\subset\Pi(M_k)$ the sets $\Pi(m_k)$ have the finite intersection property. Hence there is $p\varepsilon R^C$ belonging to all of them. Such a p defines the desired support.

We leave it to the reader to verify that if $\overline{z}$ is strongly non-tight then $p_c>0$ for all $c\varepsilon C$. ●

We conclude this section and the chapter by a short treatment of a problem which , though related to our main topic, efficiency, will not be needed in the sequel, namely the minimum wealth problem. Theorem

7.20 ensures that consumptions are expenditure minimizing in the sense that a consumption giving at least the same satisfaction can not be bought at a lower expenditure. Is it also true that preferred consumptions will cost more?

For consumers with possibly incomplete preferences, but open preferred sets, we can prove a result in this direction when there is conflict of interest at the allocation; a concept that will now be defined.

DEFINITION 7.21. Let (Ψ,C,M) be an economy over a market structure and $\bar{z}=((\bar{x}_i)_{i\epsilon I^m},(\bar{y}_j)_{j\epsilon J^m})_{m\epsilon M}$ a feasible allocation. Let

$$I_o=\{i\epsilon I \mid P_i(\bar{x}_i) \text{ is open in } X_i\}.$$

There is <u>conflict of interest at</u> $\bar{z}$ if for all $\bar{h}\epsilon I_o$ there exists a finite subset $F\subset I\cup J$ with $\bar{h}\epsilon F$ and satisfying

(i) for some $h''\epsilon F\cap I_o$ and $z\epsilon -R^{m(h'')}_{++}$

$$\Sigma_{i\epsilon F\cap I_o}\bar{x}_i - z\epsilon\Sigma_{i\epsilon F\cap I_o}X_i$$

(ii) there is a feasible allocation $z=((x_i)_{i\epsilon I^m},(y_j)_{j\epsilon J^m})_{m\epsilon M}$ such that

$$x_i\epsilon \mathrm{cl}P_i(\bar{x}_i) \quad \text{for } i\epsilon(F\cap I)\setminus\{\bar{h}\}$$

$$x_i\epsilon P_i(\bar{x}_i) \quad \text{for } i\epsilon(F\cap I_o)\setminus\{\bar{h}\}$$

$$x_h=\bar{x}_h,\ y_h=\bar{y}_h \quad \text{for } h\epsilon(I\cup J)\setminus F. \quad \bullet$$

The terminology is motivated by (ii) of Definition 7.21 which states intuitively, that consumer $\bar{h}$ is able to give up something which may, directly or indirectly, be used to increase the sastisfaction of the other consumers (in $F\cap I_o$). Similar assumptions are encountered in the literature under the term "irreducibility",see e.g Arrow and Hahn[1971] Bergstrom[1976],Borglin [1976],Borglin and Keiding [1981], Debreu[1962], Hildenbrand [1974].

THEOREM 7.22. Let (Ψ,C,M) be an economy over a market structure satisfying assumption 7.9, and let $\bar{z}=((\bar{x}_i)_{i\epsilon I^m},(\bar{y}_j)_{j\epsilon J^m})_{m\epsilon M}$ be a fesible allocation with support $p\epsilon R^C$. If there is conflict of interest at $\bar{z}$, then, for $i\epsilon I_o$

$$x_i\epsilon P_i(\bar{x}_i) \text{ implies } px_i>p\bar{x}_i$$

Proof: If $h\varepsilon I_o$ is a consumer for whom the minimum wealth situation does not occur then there is $x_h\varepsilon X_h$ such that $px_h<p\bar{x}_h$. Assume that for some $x_h'\varepsilon X_h$ we have

$$x_h'\varepsilon P_h(\bar{x}_h) \text{ and } px_h'\leq p\bar{x}_h$$

Let $x_h(\lambda)=\lambda x_h+(1-\lambda)x_h'$, then $px_h(\lambda)=p(\lambda x_h+(1-\lambda)x_h')<p\bar{x}_h$ for $0<\lambda\leq 1$ and, for λ sufficiently close to 0, we have $x_h(\lambda)\varepsilon P_h(\bar{x}_h)$. Thus p can not be a support to $\bar{z}$ and this contradiction shows that if the minimum wealth situation does not occur for a consumer $h\varepsilon I_o$ then

$$x_h\varepsilon P_h(\bar{x}_h) \text{ implies } px_h>p\bar{x}_h$$

Assume that the minimum wealth situation occurs for $\bar{h}\varepsilon I_o$. Let F be a finite subset of I as in Definition 7.21. Using (i) of Definition 7.21 we get consumptions x_i, $i\varepsilon F\cap I_o$, and a $z\varepsilon -R_{++}^{m(h")}$ for some $h"\varepsilon F\cap I_o$ such that

$$\Sigma_{i\varepsilon F\cap I_o}\bar{x}_i-z=\Sigma_{i\varepsilon F\cap I_o}x_i \text{ and } x_i\varepsilon X_i \text{ for } i\varepsilon F\cap I_o$$

Since $p_c>0$ for some $c\varepsilon m(h")$ we have for some consumer $h'\varepsilon F\cap I_o$

$$px_{h'}<p\bar{x}_{h'}$$

Hence the minimum wealth situation does not occur for consumer h'.

Let $z=((x_i)_{i\varepsilon I^m},(y_j)_{j\varepsilon J^m})_{m\varepsilon M}$ be an allocation having the properties of (ii) in Definition 7.21. Since $h'\neq\bar{h}$ and z differs from $\bar{z}$ only for finitely many i, j we get

$$p\Sigma_{i\varepsilon I\setminus\{\bar{h}\}}(x_i-\bar{x}_i)>0$$

On the other hand , feasibility and profit maximization implies

$$p\Sigma_{i\varepsilon I}(x_i-\bar{x}_i)=p\left[(\Sigma_{j\varepsilon J}y_j+\Sigma_{i\varepsilon I}\omega_i)-(\Sigma_{j\varepsilon J}\bar{y}_j+\Sigma_{i\varepsilon I}\omega_i)\right]=p\Sigma_{j\varepsilon F}(y_j-\bar{y}_j) \leq 0$$

It follows that $p(x_{\bar{h}}-\bar{x}_{\bar{h}})<0$ which shows that the minimum wealth situation does not occur for $\bar{h}$. ●

CHAPTER 8: EFFICIENCY CRITERIA FOR MANY-GOODS MODELS

In the preceding Chapters 6 and 7 we have presented a generalization to arbitrary commodity spaces and market structures of one of the central concepts in Part I, namely reduced models. In this chapter we proceed to study efficiency of such models.

One purpose is to show how the general efficiency criterion from Chapter 3, under suitable assumptions , can be generalized to the many-goods case. That such a generalization is possible is not surprising. However, to obtain it we need an additional closedness assumption.

To state a criterion parallelling that given in Theorem 3.9 we define, for the many-goods case, a composition o on sets of a reduced model over a canonical market structure. Since the assumption of closedness is crucial, Section 8.3 is devoted to giving a general condition sufficient for closedness as well as two examples illustrating the difficulty which occurs.

8.1. THE PARETO IMPROVEMENTS IN A REDUCED MODEL

Let $(S^m, S^m_*)_{m \in M}$ be a generalized reduced model. If (C,M) is a market structure on a finite set C, and we keep in mind the interpretation of S^m_* as changes in the position of some consumer(s) which result in something preferred, then the set of Pareto improvements can be defined straightforwardly as

$$\bigcup_{\bar{m} \in M} [S^{\bar{m}}_* + \Sigma_{m \in M \setminus \{\bar{m}\}} S^m]$$

This is essentially the construction to be found in the proof of the Second Fundamental Theorem of Welfare Economics (cf. Debreu [1959], Theorem 6.4, Malinvaud [1972] p.104)

What is important in our present situation is that this set of Pareto improvements makes perfect sense also in the case of a general market structure (C,M). This is due to the fact that summation of sets over infinite index sets can be carried out provided the carriers of these sets do not overlap in too complicated a way.

Let (C,M) be a market structure, $\{A^m | m \in M\}$ a family of sets, indexed by M, such that $A^m \subset R^m$, $m \in M$. We define the sum $\Sigma_{m \in M} A^m$ by

$$\Sigma_{m \in M} A^m = \{x \in R^C \,|\, x = \Sigma_{m \in M} x^m,\ x^m \in A^m,\ m \in M\}$$

Since for each $c \in C$, the set of markets m such that $c \in m$ is finite , the summations $x = \Sigma_{m \in M} x^m$ are meaningful, and so is the set $\Sigma_{m \in M} A^m$.

Returning to the generalized reduced models, we note that

$$\bigcup_{\bar{m} \in M} [S^{\bar{m}}_* + \Sigma_{m \in M \setminus \{\bar{m}\}} S^m]$$

is a well-defined subset of R^C.

The following theorem is an immediate consequence of the definitions:

THEOREM 8.1. A generalized reduced model $(S^m, S^m_*)_{m\epsilon M}$ is efficient if and only if

$$0 \notin \bigcup_{\bar{m}\epsilon M} [S^{\bar{m}}_* + \Sigma_{m\epsilon M\setminus\{\bar{m}\}} S^m]. \bullet$$

Comparing Theorem 8.1 with Theorem 3.9 we notice some important differences.

First of all, the efficiency criterion given in Theorem 3.9 involves only the finite submodels of the original model. This in turn made it useful as a starting point for deriving parametric efficiency criteria. Secondly, in the proof of Theorem 3.9, when the model was inefficient we were able to define an improvement inductively by considering the sets $(S_t o \ldots o S_T)_{T\epsilon N}, T \geq t, t\epsilon N$. These sets had a nice interpretation in the one-good case; they showed the substitution possibilities between a given market and markets far away.

We shall see that it is possible also in the many-goods case to give a criterion involving only the finite submodels of the given model and that using the canonical market structure, the interpretation of the composition of sets is retained. To state the criterion we need to define the composition of sets in the many-goods case. This will be done in the next section.

We conclude this section with a simplified version of Theorem 8.1 for reduced models where $S^m + \dot{R}^m \subset \text{int} S^m$.

COROLLARY. Let $\Sigma = (S^m)_{m\epsilon M}$ be a reduced model with $S^m + \dot{R}^m_+ \subset \text{int} S^m$ for $m\epsilon M$ and let $\bar{m}$ be an arbitrary market in M. Σ is efficient if and only if

$$0 \notin [(S^{\bar{m}} + \dot{R}^{\bar{m}}_+) + \Sigma_{m\epsilon M\setminus\{\bar{m}\}} S^m]$$

<u>Proof</u>: The "only if" part is trivial. Assume that Σ is not efficient. Then, by Theorem 8.1, for some $m'\epsilon M$,

$$0 \epsilon [(S^{m'} + \dot{R}^{m'}_+) + \Sigma_{m\epsilon M\setminus\{m'\}} S^m]$$

showing that there is an improvement $(\xi^m)_{m\epsilon M}$ such that $\xi^{m'} \epsilon S^{m'} + \dot{R}^{m'}_+$.

Let $\bar{M}$ be a connected set of markets containing m' and $\bar{m}$ such that $|\bar{M}|$ is minimal. If $|\bar{M}|=2$ the conclusion follows immediately using the monotonicity properties of the sets S^m, $m\epsilon M$. An induction on $|\bar{M}|$ gives the conclusion in the general case. $\bullet$

8.2. THE COMPOSITION OF SETS

In Chapter 3 we defined the composition of sets in R^2. Although the generalization to the many-goods case is straightforward additional notation is needed in order to treat explicitly the properties of o, in particular, to show that o is associative. The composition o of sets can be generalized to arbitrary reduced models only when the market structure is canonical.

Let $M=\{m_o,m_1,...\}$ be a canonical market structure and define the first and last part of market m_i by

$$m_{iF}=m_{i-1}\cap m_i \quad \text{for } i\epsilon N$$

$$m_{iL}=m_i\cap m_{i+1} \quad \text{for } i\epsilon N\cup\{0\}$$

Note that market m_i may contain some goods which do not occur in any other market. Since there will be no gain in simplicity of notation or otherwise we will not exlude the possibility of such internal goods. Market m_o , of course, has empty first part.

DEFINITION 8.2. Let $M=\{m_o,m_1,...\}$ be a canonical market structure and $(Z^m)_{m\epsilon M}$ a collection of sets such that $Z^m\subset R^m$. Define the <u>composition of</u> Z^{m_i} <u>and</u> $Z^{m_{i+1}}$ by

$$Z^{m_i}oZ^{m_{i+1}}=\{z\epsilon R^{m_i\cup m_{i+1}}\mid z^i\epsilon Z^{m_i},\ z^{i+1}\epsilon Z^{m_{i+1}} \text{ with } z_c^i+z_c^{i+1}=0 \text{ for } c\notin m_{i+1F}\cup m_{iL}\}. \bullet$$

The composition o is defined with respect to a given market structure; this dependence of o on M might be indicated by the more tedious notation o_M. In this notation, we may write down the associativity property of o as

$$(Z^{m_i}o_M Z^{m_{i+1}})o_{M'}Z^{m_{i+2}}=Z^{m_i}o_{M''}(Z^{m_{i+1}}o_M Z^{m_{i+2}})$$

where

$$M' = \{m_o, m_2, \ldots, m_{i-1}, m_i \cup m_{i+1}, m_{i+2}, \ldots\}$$

$$M'' = \{m_o, m_2, \ldots m_i, m_{i+1} \cup m_{i+2}, m_{i+3}, \ldots\}$$

(The verification of the details are left to the reader). Pursuing this formalism a little further , we note that $o_{M'}$, where M' runs over all the canonical market structures on C, defines a partial semigroup structure on the family of all sets $Z \subset R^m$, where $m \in M'$ for some M'. Since this partial semigroup structure is not particularly useful in the following, no further reference will be made to it. Also we drop the index M on the composition o since in all future applications, the market structure involved will be clear from the context.

Let $(S^m)_{m \in M}$ be a reduced model and M a canonical market structure. In particular, $\Sigma_{m \in F} S^m$ is closed for F a finite subset of M. We note that

$$S^{m_i} o S^{m_{i+1}} = \{z \in R^{m_i \cup m_{i+1}} \mid z = z^{m_i} + z^{m_{i+1}} \text{ for some } z^{m_i} \in S^{m_i} \text{ and } z^{m_{i+1}} \in S^{m_{i+1}}\} \cap \cap \{z \in R^{m_i \cup m_{i+1}} \mid z_c = 0 \text{ for } c \notin m_{iL} \cup m_{i+1F}\}$$

Since the right hand member of this relation is the intersection of $S^{m_i} + S^{m_{i+1}}$, a closed set, with another closed set we conclude that $S^{m_i} + S^{m_{i+1}}$ is closed.

In Section 3.2 we gave an interpretation of the composition of sets in one-good reduced models. This interpretation carries over to many-goods models; the set

$$\{z \in S^m \mid z_c = 0 \text{ for } c \notin m_F \cup m_L\}$$

gives the possibilities of substitution between the first and last part of market m and $S^{m_i} o \ldots o S^{m_T}$ for $T \geq i$ shows the possibilities of substitution between the goods in m_{iF} and m_{TL}.

8.3. A GENERAL EFFICIENCY CRITERION

In this section we shall move on from the abstract efficiency criterion of Theorem 8.1 to another version involving only finite submodels. We start with a rather technical consideration.

In the proof of Theorem 3.9 we were able to define an improvement inductively. We were also able to get along without any closedness assumption on the total model. As will be seen below, for the many-goods case, an improvement can not be defined inductively. This being the case, we have to introduce some kind of closedness assumption on the total model. A closer investigation of one-good reduced models reveals the difficulty encountered. Let $(\xi_t)_{t\in N}$ be an improvement with $\xi_1>0$ for a weakly efficient one-good reduced model. Then, for each T,

$$\Sigma_{t=1}^{T}(-\xi_t,\xi_{t+1})=(-\xi_1,\xi_2)+(-\xi_2,\xi_3)+\ldots+(-\xi_{T-1},\xi_T)\in\Sigma_{t=1}^{T}S_t\subset\Sigma_{t=1}^{\infty}S_t$$

(Here $(-\xi_t,\xi_{t+1})$ and similar expressions should be interpreted as vectors in R^N).

Thus the sequence ζ^T in R^N defined by

$$\zeta^T=\Sigma_{t=1}^{T}((-\xi_t,\xi_{t+1})=(-\xi_1,0,\ldots,0,\xi_T,0\ldots),\ T\in N$$

belongs to $\Sigma_{t=1}^{T}S_t$ and converges pointwise to $(-\xi_1,0,0,\ldots)$.

This reasoning suggests that it is the closedness of $\Sigma_{t=1}^{T}S_t$ which is essential when we want to conclude about inefficiency of the reduced model from the existence of finite sequences satisfying the relevant properties of an improvement up to date T.

On the other hand the following example shows that $\Sigma_{t=1}^{T}S^{mt}$ will, even in the one-good case, in general not be closed.

EXAMPLE 8.3. Let $(S^{mt})_{t\in N\cup\{0\}}$ be a one-good reduced model defined as in Example 7.5 and let

$$S^{mo}=\{(x_o,y_1)\in R^{mo}\,|\,(x_o,y_1)\in R_+^{mo}\}$$

$$S^{m1}=\{(x_1,y_2)\in R^{m1}\,|\,x_1+y_2-x_1y_2\geq 0,\ x_1+y_2\geq 0\}$$

$$S^{mt}=\{(x_t,y_{t+1})\in R^{mt}\,|\,x_t+y_{t+1}\geq 0\}$$

Let $((x_t^n,y_{t+1}^n)_{t\in N})_{n\in N}$ be a sequence in R^N such that

$$(x_t^n,y_{t+1}^n)\in S^{mt}\text{ for } n\in N,\ t\in N$$

$$y_t^n+x_t^n=0 \text{ for } n\epsilon N,\ t\epsilon N$$

Then, for each $n\epsilon N$,

$$\Sigma_{t=1}^{\infty}(x_t^n,y_{t+1}^n)=(x_1^n,0,0,\ldots)\epsilon\Sigma_{t=1}^{\infty}S^{mt}$$

However, if the sequence is chosen so that $x_1^n \longrightarrow -1$ as $n \longrightarrow \infty$ the pointwise limit $(-1,0,0,\ldots)$ does not belong to $\Sigma_{t=1}^{\infty}S^{mt}$. Hence $\Sigma_{t=1}^{\infty}S^{mt}$ (and $\Sigma_{t=0}^{\infty}S^{mt}$) is not closed in the product topology. ●

Since we were able to prove Theorem 3.9 without an assumption of closedness for $\Sigma_{t=1}^{\infty}S^{mt}$ it is reasonable to expect that a weaker closedness condition will suffice. In order to state such a weaker condition we need the following concept.

Let C be a set of commodity labels and A a subset of C. For B, a subset of R^C, we define the <u>projection</u> of B on R^A by

$$pr_A B=\{z\epsilon R^A \mid \exists\ z'\epsilon B,\ z'_c=z_c \text{ for } c\epsilon A\}$$

DEFINITION 8.4. Let $\Sigma=(S^m)_{m\epsilon M}$ be a reduced model. Σ is <u>closed</u> if $\Sigma_{m\epsilon M}S^m$ is closed and Σ is <u>pseudo-closed</u> if for each $\overline{m}\epsilon M$ and for each $x^{\overline{m}}\epsilon R^{\overline{m}}$, the set

$$\Sigma_{m\epsilon M\setminus\{\overline{m}\}}S^m \cap \{z\epsilon R^C \mid z_c=x_c^{\overline{m}} \text{ for } c\epsilon\overline{m}\}$$

is closed in R^C. A generalized reduced model $(S^m,S_*^m)_{m\epsilon M}$ is <u>closed</u> (<u>pseudo-closed</u>) if $(S^m)_{m\epsilon M}$ is closed (pseudo-closed). ●

Thus, for pseudo-closedness of a reduced model we do not demand that the whole of $\Sigma_{m\epsilon M}S^m$ be closed. Rather, the fibre of $\Sigma_{m\epsilon M}S^m$ over each $x^{\overline{m}}\epsilon R^{\overline{m}}$, that is to say, the set of all elements in $\Sigma_{m\epsilon M}S^m$ which project into $x^{\overline{m}}$ should be closed for any market $\overline{m}$.

EXAMPLE 8.3.(continued) If the sequence $((x_t^n,y_{t+1}^n)_{t\epsilon N})_{n\epsilon N}$ is such that $x_1^n=\overline{x}_1$ for $n\epsilon N$ and some $\overline{x}_1$ satisfying

$$\overline{x}_1\epsilon pr_{m_0}\Sigma_{m\epsilon M\setminus\{m_0\}}S^m$$

then the limit $(\overline{x}_1,0,0,\ldots)$ belongs to $\Sigma_{t=1}^{\infty}S^{mt}$. ●

Of course, if $\Sigma=(S^m)_{m\varepsilon M}$ is a reduced model such that $\Sigma_{m\varepsilon M}S^m$ is closed then Σ is pseudo-closed. Although pseudo-closedness seems to be the relevant property allowing us to generalize the results of Chapter 3 we would expect a large class of models to satisfy the stronger condition of closedness. Therefore we discuss, in the next section, assumptions which ensure closedness. There we also give an example to show that not even pseudo-closedness can be derived from the other assumptions that we have used previously.

Before stating a general criterion for efficiency of generalized reduced models we note the following difference to one-good models. For a one-good model $(S^{m_t})_{t\varepsilon N\cup\{0\}}$, as defined in Example 7.5, we always have the "preferred set" for a market m_t equal to $S^{m_t}+\dot{R}^{m_t}$ so that

$$pr_{m_t\cap m_{t+1}}(S^{m_t}\cap\dot{R}^{m_t\cap m_{t+1}})=\dot{R}_+^{m_t\cap m_{t+1}}$$

and this latter set may be identified with $\dot{R}_+$.

This independence of the "preferred set" made possible a simple formulation of the efficiency criterion for one-goods models. For many-goods models we have to be more explicit about the "preferred set" of a market.

THEOREM 8.5. Let $\Sigma_G=(S^m,S_*^m)_{m\varepsilon M}$ be a pseudo-closed generalized reduced over a canonical market structure. Then Σ_G is efficient if and only if

$$(1)\qquad \bigcup_{i=0}^{\infty}[[pr_{m_i}(S^{m_0}\circ\ldots S^{m_{i-1}})+\cap_{T\geq i+1}pr_{m_i}(S^{m_{i+1}}\circ\ldots\circ S^{m_T})]\cap(-S_*^{m_i})]=\phi. \quad\bullet$$

Though complicated-looking, the expression (1) is a rather exact counterpart of the efficiency criterion in Theorem 3.9 for the one-good model. It says thet there is no market m_i such that an improvement can be carried out through the combined effect of trade-offs in markets "before" and markets "after" the given one. We note in passing that the fact that the market structure is canonical is exploited in the very statement of the theorem.

Proof of Theorem 8.5: Assume that Σ_G is inefficient and let $(\xi^m)_{m\varepsilon M}$ be an improvement; choose $i\varepsilon N\cup\{0\}$ such that $\xi^{m_i}\varepsilon S_*^{m_i}$. Using Theorem 8.1 we get that

$$-\xi^{m_i}=\Sigma_{j\neq i}\xi^{m_j}$$

so that

$$pr_{m_i}(\Sigma_{j\neq i}\xi^{m_j})=-\xi^{m_i}$$

Hence

$$(2)\quad [pr_{m_i}(S^{m_0}\circ\ldots\circ S^{m_{i-1}})+\cap_{T\geq i+1}(S^{m_{i+1}}\circ\ldots\circ S^{m_T}]\cap(-S^{m_i}_{*})\neq\phi$$

Thus (1) implies that Σ_G is efficient.

Conversely, suppose that (1) is false, so that (2) holds for some i. Choose $-\xi^{m_i}$ belonging to the intersection in (2). Then by definition there is $\xi^{m_i}\varepsilon S^{m_i}$ and for every $T\geq i+1$ a sequence $(\xi_T^{m_i})_{i=0}^{T}$ with

$$\xi_T^{m_i}=\xi^{m_i},\ \xi_T^{m_j}\varepsilon S^{m_j},\ j\leq T$$

and such that the vector $\Sigma_{j=0}^{T}\xi_T^{m_j}$ in R^C has non-zero coordinates only for $c\varepsilon m_T$

It follows that the sequence $(\Sigma_{j\neq i}\xi_T^{m_j})_{T\geq i+1}$ of vectors in $\Sigma_{j\neq i}S^{m_i}\subset R^C$ converges pointwise to the vector $-\xi^{m_i}(\varepsilon R^C)$. By pseudo-closedness, $-\xi^{m_i}\varepsilon_{j\neq i}S^{m_j}$, and we conclude that there is an improvement for Σ_G. Thus efficiency of Σ_G implies (1). ●

Clearly, Theorem 8.5 as formulated above is rather less general than the other results of this and previous chapters since it holds only for canonical market structures. It should be recalled at this point that arbitrary reduced models may be transformed into models over canonical market structures by a suitable morphism, cf. Theorem 6.7. The point to be made, however, is that this implies that certain operations of summation of sets S^m and S^m_*, are performed, thereby transforming the originally given sets in a way which in applications may be difficult to evaluate. In view of this we have chosen the present formulation of Theorem 8.5 rather than a more general version.

We can state a simplified version of Theorem 8.5 for sufficiently regular reduced models.

COROLLARY. Let $\Sigma=(S^m)_{m\in M}$ be a pseudo-closed reduced model over a canonical market structure such that

$$S^m+\dot{R}^m_+\subset \text{int} S^m$$

Σ is efficient if and only if

$$(\cap_{T=1}^{\infty} pr_{m_o} S^{m_1}o\dots S^{m_T})\cap(-pr_{m_o}(S^{m_o}+\dot{R}^{m_o}_+))=\phi$$

Proof: This follows easily from Theorem 8.5 together with the Corollary of Theorem 8.1. ●

8.4.* CLOSED AND PSEUDO-CLOSED REDUCED MODELS

Although pseudo-closedness was sufficient in order to derive a general criterion for efficiency one would perhaps expect reduced models derived from economies, satisfying reasonable assumptions, to satisfy the stronger assumption of closedness. Section 8.4.1 is devoted to a generalization of the "asymptotic cone" condition used in finite economies. It turns out that, loosely speaking, if production possibilities are such that each market contains some factor essential for producing output on that market so that, on the other hand, commodities outside the market are of limited use for producing output on that market then the corresponding reduced model will be closed. We begin by abstracting from the economic problem at hand and first state a condition for an infinite sum of closed sets, each contained in a finite dimensional cooordinate subspace, to be closed. Then we go on in Section 8.4.2 to show that a many-goods generation model, satisfying reasonable assumptions will be closed.

In Section 8.4.2 we also give examples to show that the derived condition is not necessary for closedness and that not even pseudo-closedness can be derived form the rest of our assumptions.

8.4.1. A Sufficient Condition for Closedness

Recall that for a set $Q \subset R^F$, F a finite subset of C, the asymptotic cone of Q, denoted $A[Q]$ is

$$A[Q] = \{z \in R^F \mid \exists z' \text{ with } z = \lambda z',\ \lambda \geq 0 \text{ and there exists a sequence } z^k \in Q,\ k \in N, \text{ such that } \|z^k\| \longrightarrow \infty \text{ and } z^k/\|z^k\| \longrightarrow z'\}$$

If $Q \subset R^F$ is a closed, convex set owning 0, then $A[Q] \subset Q$. If Q^i, $i=1,\dots,k$, are closed , convex sets in R^F and $z^i \in A[Q^i]$, $i=1,\dots,k$, and $\Sigma_{i=1}^k z^i = 0$ implies $z^i = 0$ for $i=1,\dots,k$ then the asymptotic cones of Q^i, $i=1,\dots,k$, are positively semi-independent. If this is the case then $\Sigma_{i=1}^k Q^i$ is closed.(Cf. Debreu[1959])

In the finite dimensional case the assumption that the asymptotic cones of the individual production sets are positively semi-independent prevents the possibility of "reversible production at arbitrary scale". More precisely, when $0 \in Y$, there can not be $y \neq 0$ belonging to the total production set such that $\lambda y \in Y$ for $\lambda \in R$

To treat the infinite dimensional case we make the following:

DEFINITION 8.6. Let (C,M) be a market structure and $(Q^m)_{m \in M}$ a family of sets such that $Q^m \subset R^m$, $m \in M$. $(A[Q^m])_{m \in M}$ are <u>semi-independent</u> on B,a finite subset of C, if $z^m \in A[Q^m]$, $m \in M$, and $\Sigma_{m \in M} pr_B z^m = 0$ implies $pr_B z^m = 0$ for $m \in M$.

If there is an increasing sequence of finite sets, $(B_k)_{k \in N}$, with $\cup_{k \in N} B_k = C$ such that $(A[Q^m])_{m \in M}$ are semi-independent on each B_k then $(A[Q^m])_{m \in M}$ are <u>finitely semi-independent</u>. ●

To give an interpretation of this concept, assume that (C,M) is a market structure and that $(Q^m)_{m \in M}$ is a family of closed, convex sets such that $0 \in Q^m$ for $m \in M$ and that $(A[Q^m])_{m \in M}$ are not finitely semi-independent.Then, for any sequence $(B_k)_{k \in N}$ satisfying the properties of Definition 8.6, there will be at least one B_k such that $(A[Q^m])_{m \in M}$ are not semi-independent on B_k. Put $B_k = B$. There are $\bar{z}^m \in A[\bar{Q}^m]$, $m \in M$, such that

(i) $z^m \in A[Q^m] \subset Q^m$

(ii) $\Sigma_{m\in M} pr_B z^m = 0$ and $pr_B z^{\bar{m}} \neq 0$ for some $\bar{m}\in M$

By (ii) we then have

$$pr_B z^{\bar{m}} = -\Sigma_{m\in M\setminus\bar{m}} pr_B z_m \in -\Sigma_{m\in M} pr_B Q^m$$

Since, for $m\in M$, $A[Q^m]$ is a cone and $A[Q^m]\subset Q^m$ we get

$$\lambda pr_B z^{\bar{m}} \in (\Sigma_{m\in M} pr_B Q^m) \cap (-\Sigma_{m\in M} pr_B Q^m) \text{ for } \lambda\in R$$

Hence if the sets Q^m, $m\in M$, are production sets then from the relation above we get that there are productions y, $y'\in\Sigma_{m\in M} Q^m$ with

$$pr_B y \neq 0,\ pr_B y' \neq 0,\ \lambda y \in \Sigma_{m\in M} Q^m \text{ and } \lambda y' \in \Sigma_{m\in M} Q^m \text{ for } \lambda\in R$$

and

$$pr_B y = -pr_B y'$$

Hence, by varying the inputs of goods not in B, we obtain productions whose restrictions to B, in a sense, show "the possibility of reversible production at arbitrary scale" for the goods in B.

We can now state:

THEOREM 8.7. Let M be a market structure and $(Q^m)_{m\in M}$ a family of closed sets with $Q^m \subset R^m$. If $(A[Q^m])_{m\in M}$ are finitely semi-independent then $\Sigma_{m\in M} Q^m$ is closed.

<u>Proof</u>: We want to show that $Q=\Sigma_{m\in M} Q^m$ is closed, or equivalently, that if $y^k\in Q$, $k\in N$, and y^k converges pointwise to y then $y\in Q$.

Let $y^k\in Q$, $k\in N$, converge to y. Then there are $y^{m,k}\in Q^m$, $m\in M$, $k\in N$ such that $\Sigma_{m\in M} y^{m,k} = y^k$ converges to y.

<u>Case</u> 1. For each $m\in M$ there is a subsequence $N_m \subset N$ such that $(y^{m,k})_{k\in N_m}$ is bounded in norm.

Using Cantor's diagonal procedure we can then find a subsequence N_o such that, for $m\epsilon M$, $(y^{m,k})_{k\epsilon N_o}$ converges to, say, y^m. Since Q^m is closed for each m; $y^m \epsilon Q^m$ and $y=\Sigma_{m\epsilon M} y^m \epsilon \Sigma_{m\epsilon M} Q^m$.

Case 2. For some $\bar{m}\epsilon M$ and any subsequence $N_o \subset N$ $(y^{\bar{m},k})_{k\epsilon N_o}$ is such that $(\|y^{\bar{m},k}\|)_{k\epsilon N_o}$ is not bounded. By assumption, there is a sequence of sets $(B_k)_{k\epsilon N}$, $\cup_{k=1}^{\infty} B_k = C$ such that $(A[Q^m])_{m\epsilon M}$ are semi-independent on B_k, $k\epsilon N$. Choose k so that $m \subset B_k$, let

$$D=\{m\epsilon M \mid m\cap B_k \neq \phi\}$$

and let

$$\sigma_k = \Sigma_{m\epsilon D} \|y^{m,k}\|$$

Then

$$\frac{y_k}{\sigma_k} = \Sigma_{m\epsilon D} \frac{\|y^{m,k}\|}{\sigma_k} \frac{y^{m,k}}{\|y^{m,k}\|} + \Sigma_{m\epsilon M \backslash D} \frac{\|y^{m,k}\|}{\sigma_k} \frac{y^{m,k}}{\|y^{m,k}\|}$$

Without loss of generality we may assume that, for $m\epsilon D$, $\|y^{m,k}\|/\sigma_k$ converges to α^m and $y^{m,k}/\|y^{m,k}\|$ converges to y^m.

Hence, for $m\epsilon D$,

$$\frac{\|y^{m,k}\|}{\sigma_k} \frac{y^{m,k}}{\|y^{m,k}\|} \longrightarrow \alpha^m y^m \epsilon A[Q^m]$$

Furthermore, $y^{\bar{m}} \neq 0$ and $\Sigma_{m\epsilon D} \alpha^m = 1$, $\alpha^m \geq 1$ for $m\epsilon D$.

Since y^k converges to y we have

$$\lim pr_{B_k} \frac{y^k}{\sigma_k} = 0 = \lim pr_{B_k} (\Sigma_{m\epsilon M} \frac{\|y^{m,k}\|}{\sigma_k} \frac{y^{m,k}}{\|y^{m,k}\|})$$

and as pr_{B_k} is a continuous linear map, it commutes with the limit operation. Whence

$$\Sigma_{m\epsilon M} \alpha^m pr_{B_k} y^m = 0$$

Since $pr_{B_k} y^{\bar{m}} = y^{\bar{m}} \neq 0$ this contradicts that $(A[Q^m])_{m\epsilon M}$ are semi-independent on B_k and hence, by Definition 8.6, that they are finitely semi-independent.

It follows that Case 2 can not occur and that $\Sigma_{m\epsilon M} Q^m$ is closed. ●

COROLLARY. Let $\Sigma=(S^m)_{m\epsilon M}$ be a reduced model such that $(A[S^m])_{m\epsilon M}$ are finitely semi-independent then $\Sigma_{m\epsilon M}S^m$ is closed. ●

8.4.2. Applications to Generation Models and Two examples.

The usual reason advanced for "the impossibility of irreversible production" is that goods are dated. Applied to a generation economy this implies that the producers acting on market m can use the goods in m_F only as inputs and that it is not possible to use only goods in m_L to produce goods in m_L. Hence with no input of goods from m_F, possible output of goods in m_L should at least be bounded from above. This is made precise by the following:

THEOREM 8.8. Let (Ψ,C,M) be an economy over a canonical market structure with $M=\{m_o,m_1,...\}$ satisfying Assumption 7.9. For a feasible allocation $z=((\bar{x}_i)_{i\epsilon I^m},(\bar{y}_j)_{j\epsilon J^m})_{m\epsilon M}$ let $(S^m,S^m_*)_{m\epsilon M}$ be defined by (3) and (8) of Chapter 7. Assume that $(S^m,S^m_*)_{m\epsilon M}$ is a generalized reduced model and that

(i) X_i has a lower bound for $i\epsilon I$

(ii) for each $m\epsilon M\backslash m_o$ and $y\epsilon A[\Sigma_{j\epsilon J^m}Y_j]$; if $pr_{m_F}y=0$ then $pr_{m\backslash m_F}y\epsilon -R_+^{m\backslash m_F}$

and

for $y\epsilon A[\Sigma_{j\epsilon J^{m_o}}Y_j]$; if $pr_{m_o\backslash m_{oL}}y=0$ or $m_o\backslash m_{oL}=\phi$ then $pr_{m_{oL}}y\epsilon -R_+^{m_{oL}}$

Then Σ is closed.

<u>Proof</u>: Recall that

$$S^m=\Sigma_{i\epsilon I^m}\, cl(P_i(\bar{x}_i)-\bar{x}_i)-\Sigma_{j\epsilon J^m}(Y_j-\bar{y}_j)$$

Since translation does not alter the asymptotic cone of a set we get, using (i),

$$A[S^m]= A[\Sigma_{i\epsilon I^m}\, cl(P_i(\bar{x}_i)-\bar{x}_i)-\Sigma_{j\epsilon J^m}\,(Y_j-\bar{y}_j)]\subset$$

$$\subset A[\Sigma_{i\epsilon I^m}\, X_i-\Sigma_{j\epsilon J^m}\, Y_j]\subset A[R^m-\Sigma_{j\epsilon J^m}\, Y_j]\subset$$

$$\subset A[-((\Sigma_{j\epsilon J^m}\, Y_j)-R^m)\subset -A[\Sigma_{j\epsilon J^m}\, Y_j]$$

To prove that the sets S^m, $m\epsilon M$, are finitely semi-independent choose $B_k=m_o\cup m_1\cup\ldots\cup m_k$, $k\epsilon N$.

Consider some $k\epsilon N$ and $(z^m)_{m\epsilon M}$ with $z^m\epsilon A[S^m]$ such that

$$\Sigma_{m\epsilon M}pr_{B_k}z^m=0$$

Since the market structure is canonical we get by (ii)

$$pr_{m_o\setminus m_{oL}}z^{m_o}=0 \text{ or } m_o\setminus m_{oL}=\phi$$

Hence, since $z^{m_o}\epsilon -A[\Sigma_{j\epsilon J^{m_o}}Y_j]$

$$pr_{m_{oL}}z^{m_o}\epsilon R_+^{m_{oL}}$$

The same reasoning applied to z^{m_1} shows that

$$pr_{m_{1F}}z^{m_1}\epsilon R_+^{m_{1F}}$$

Since $m_{oL}=m_{1F}$ and $\Sigma_{m\epsilon M}pr_{B_k}z^m=0$ we get

$$pr_{m_{1F}}(z^{m_o}+z^{m_1})=0$$

Hence $z^{m_o}=0$ and, by induction it follows that

$$z^{m_i}=0 \text{ for } i=0,1,2,\ldots,k-1 \text{ and } z^{m_{kF}}=0$$

which implies

$$pr_{B_k}z^{m_i}=0 \text{ for } i\epsilon N\cup\{0\}$$

Hence, for any B_k,$k\epsilon N$, the sets $A[S^m]$,$m\epsilon M$ are semi-independent on B_k and thus, by Definition 8.6, finitely semi-independent. It now follows from the corollary to Theorem 8.7 that Σ is closed. ●

The following example shows that the condition for closedness is not a necessary condition.

EXAMPLE 8.9. Let $C=\{0,1,2,\ldots\}$ and let $M=\{\{i,i+1\}|i\epsilon C\}$. Interpret the commodities as being identical except for different locations (also indexed by elements of C). Assume that transportation between neighboring locations is costless in both directions, that is, we have production possibilities

$$Y^{m_i}=\{y\in R^C \mid y_c=0 \text{ for } c\notin\{i,i+1\} \text{ and } y_i+y_{i+1}\leq 0\}$$

for $m_i=\{i,i+1\}\in M$, $i\in C$.

It is easy to check that for any finite subset $F\subset M$, $A[Y^m]$, $m\in F$, are positively semi-independent but that $(A[Y^m])_{m\in M}$ are not finitely semi-independent. The same, of course, holds for the reduced model corresponding to $(Y^m)_{m\in M}$.

Consider the total production set $Y=\Sigma_{m\in M}Y^m$. Choose, for $i\in N\cup\{0\}$,

$$y_i^{m_i}=i+1$$

$$y_{i+1}^{m_i}=-(i+1)$$

Then $\Sigma_{m\in M}y^m=(1,1,1,\ldots)$ and, since Y is a cone with free disposal, $Y=R^C$. Consequently Y is closed. We also note that Y lacks efficient points. ●

To see that there exists reduced models which are not even pseudo-closed we give another example.

EXAMPLE 8.10. In the following example we work with the production sets instead of the corresponding reduced model.

Consider a model with only producers where C=N and

$$M=\{\{1\},\{1,2,3\},\ m_{2t},\ t\in N\}$$

where $m_0=\{1\}$, $m_1=\{1,2,3\}$ and $m_{2t}=\{2t,2t+1,2t+2,2t+3\}$ for $t\in N$.

The production sets are closed, convex cones with free disposal. There is one producer for each market in M, indexed by the number of the market,

$$Y_0=\{y_1^- \mid y_1^-\leq 0\}$$

$$Y_1=\{(y_1^+,y_2^-,y_3^+) \mid y_1^++y_2^-+y_3^+\leq 0\}$$

$$Y_{2t}=\{(y_{2t}^+,y_{2t+1}^-,y_{2t+2}^-,y_{2t+3}^+) \mid y_{2t+1}^-+y_{2t+3}^+\leq 0 \text{ and}$$

$$y_{2t}^+\leq a_{2t}(\min(-y_{2t+1}^-,-y_{2t+2}^-\}$$

where $a_2=3/2$, $a_{2t}<1$ and $h_t=a_2\ldots a_{2t}>1$ but h_t decreases to 1 as t tends to infinity. Varibles with superscript + (-) are always non-negative (non-positive).

The following diagrams may be helpful

Good Producer	1	2	3	4	5	6	7	8
1	y_1^+	y_2^-	y_3^+					
2		y_2^+	y_3^-	y_4^-	y_5^+			
4				y_4^+	y_5^-	y_6^-	y_7^+	

Good Producer	2t-2	2t-1	2t	2t+1	2t+2	2t+3	2t+4	2t+5
2(t-1)	y_{2t-2}^+	y_{2t-1}^-	y_{2t}^-	y_{2t+1}^+				
2t			y_{2t}^+	y_{2t+1}^-	y_{2t+2}^-	y_{2t+3}^+		
2(t+1)					y_{2t+2}^+	y_{2t+3}^-	y_{2t+4}^-	y_{2t+5}^+

The situation for producer 2t may be described as follows. Producer 2t produces good 2t for which goods 2t+1, 2t+2 are essential as inputs. The input of good 2t+1, as well as being useful in the production of good 2t, results in output of good 2t+3. Producer 2(t+1) is the sole supplier of good 2t+2 and, in turn, needs 2t+3 as input to produce this good. Good 2t+3 has to be supplied by producer 2t. On the other hand, producer 2t-1 is the sole supplier of good 2t+1 which is needed to produce good 2t. Producer 2t must thus supply the producers following him (with good 2t+3) and the producer preceding him (with good 2t). Hence there is no clear time direction of production activities.

We will now show that

$$1 \in pr_1 \Sigma_{m \in M \setminus \{m_o\}} Y^m$$

and, furthermore, exhibit a sequence contained in

$$Y'=\Sigma_{m\varepsilon M\setminus\{m_0\}}Y^m\cap\{z\varepsilon R^N|z_1=1\}$$

which converges to (1,0,0,...). However, (1,0,0,...) is not contained in Y' - and it follows that the corresponding reduced model is not pseudo-closed.

Let $T(i)=\{m_1,m_2,m_4,...,m_{2i}\}$ for $i\varepsilon N$. For each $d\geq 0$

$$(1,-(d+1),d,0,0,...)\varepsilon Y_1.$$

We will now show that for each for each $i\varepsilon N$, we can choose d so that $(1,0,...,0,-d,d,0,...)\varepsilon\Sigma_{m_t\varepsilon T(i)}Y^{m_t}$. Here -d and d refer to the amounts of commodity 2i+2 and 2i+3 respectively.

Consider first a given $i\varepsilon N$. Choose $d\geq 0$. What is the maximum amount of y_2^+ possible when $y_3^-+y_3^+\leq 0$?

Clearly

$$y_2^+\leq a_2\min(y_4^+,d)\leq a_2\min(a_4\min(y_6^+,d),d)\leq$$

$$\leq a_2\min(a_4\min(a_6\min(...(a_{2i}\min(-y_{2i+2}^-,d),d)\leq a_2a_4...a_{2i}d$$

with equality if $-y_{2i+2}^-\geq d+1$.

Thus in order to get $y_2^+\geq d+1$ we must have

$$d\geq\frac{1}{a_2a_4...a_{2i}-1}$$

Still considering a given $i\varepsilon N$, choose

$$d=\frac{1}{a_2a_4...a_{2i}-1}$$

and

$$y^{m_1}=(1,-(d+1),d,0,...)$$

$$y^{m_{2t}}=(0,...,0,h_td,-d,-\frac{h_t}{a_{2t}}d,d)$$

where $h_t=a_2a_4...a_{2t}$ as before.

Then $y^{m_1}+\Sigma_{t=1}^i y^{m_{2t}}=(1,0,...,-d,d)$.

Summing up: For each $i \in N$ we can choose d so that

$$y^i=(1,0,\dots,-d,d,\dots)\in Y^{m1}+\Sigma_{t=1}^{i}Y^{m2t}\subset Y^{m1}+\Sigma_{t=1}^{\infty}Y^{m2t}$$

We note that y^i converges pointwise to $y=(1,0,\dots)$.

Assume that $y=(1,0,\dots)\in Y^{m1}+\Sigma_{t=1}^{\infty}Y^{m2t}$. Then there are y^{m1} and y^{m2t}, $t\in N$, such that $y=y^{m1}+\Sigma_{t=1}^{\infty}y^{m2t}$ and for some $d\geq 0$

$$y^{m1}\leq(1,-(d+1),d) \text{ and } y_2^{+}\geq d+1$$

But then, as the reasoning above shows,

$$d\geq \frac{1}{a_2a_4\dots a_{2i}-1}$$

for any $i\in N$. Since $a_2a_4\dots a_{2i}$ converges to 1 as i tends to infinity this is impossible. Hence $y\notin Y^{m1}+\Sigma_{t=1}^{\infty}Y^{m2t}$. ●

CHAPTER 9: APPROXIMATIONS OF MANY-GOODS MODELS

In Chapter 4 and 5 we introduced a measure of curvature which turned out to be useful in efficiency considerations. The purpose of this chapter is to provide a generalization to many-goods models of this measure of curvature, as well as, of the results of Chapter 5.

As might be expected most of the results of Chapter 5 carry over to the many-goods case. This is so because the measure of curvature in higher dimensions to be introduced below, extending that of Part I, is essentially two-dimensional. However, a distinguishing feature of our present discussion as compared to that of Part I is that now we allow for more complicated market structures.

The measure of curvature will be defined only when markets have a first and last part which again means that we deal with a canonical market structure. However, we know from Theorem 6.7 that any market structure may be brought on this form. Hence our results apply with minor qualifications to general reduced models.

In Chapter 4 we introduced the concept of a morphism. Morphisms were useful for extending results valid for a small class of reduced models to a larger class. In Section 6.1 we have given the general notion of a morphism for many-goods models. The main difference from Chapter 4 arose from the fact that we wanted to allow for different market structures.

We shall rely on morphisms in this chapter, not only when deriving results, but also for stating desirable invariance properties for the measure of curvature.

9.1 EXAMPLES OF MORPHISMS

The task performed by morphisms as introduced in Chapter 4 was to change units of measurement of the goods in the model. Through this change of units, formally accomplished by applying a morphism, we were able to extend the results for reduced models with support $p_t=1$ to models with arbitrary supports.

Before we proceed with the main topic of this, chapter we give some examples of morphisms in the many-goods case which will be useful in the sequel. Recall that a morphism was defined as a pair (η,α), $\eta=(\gamma,\iota)$ where $\gamma:C\rightarrow C'$, $\iota:M\rightarrow M'$, $\alpha\varepsilon R^C$ and that a morphism induced a linear mapping $g_{\eta,\alpha}$ from $R^{(C)}$ to $R^{(C')}$.

EXAMPLE 9.1. Let $\gamma:C\rightarrow C$ be the identity. In this case $g_{\eta,\alpha}$ induces a <u>change in the units of measurement</u>; one old unit of good c being equal to α_c new units. ●

EXAMPLE 9.2. Let $\gamma:C\rightarrow C'$ be a bijection such that M, the market structure on C, is $\{\gamma^{-1}(m')|m'\varepsilon M'\}$ and let $\alpha=1_C$. Then (η,α) is a <u>relabelling</u> or, if C=C', a <u>permutation</u> of goods "within" markets. ●

EXAMPLE 9.3. Let (η,α) be a market morphism, so that $\alpha=1_C$, and $\{\gamma^{-1}(m')|m'\varepsilon M'\}=M$. Then γ is surjective. If γ is many-to-one, (η,α) may be said to be an <u>aggregation</u> of goods in C.

As an illustration, let $M=\{m_0,m_1,m_2,...\}$ be a canonical market structure without internal goods so that

$$m_i=(m_i\cap m_{i-1})\cup(m_i\cap m_{i+1}) \text{ for } i\in N\cup\{0\}$$

and

$$\gamma(c)=\min\{i\mid c\in m_i\}$$

Then (γ,ι) maps (C,M) to $(N\cup\{0\},M')$ where

$$M'=\{\{0\},\{0,1\},\{1,2\},\ldots\}$$

by aggregation of the goods in $m_i\cap m_{i-1}$ for $i\in N$. ●

9.2. AXIOMS FOR AN APPROXIMATING FAMILY

We now proceed to consider approximating families relative to a given commodity space. Since the set of commodity labels may be chosen in many different ways, it is important that the sets in the approximating family posess certain invariance properties under changes of commodity labels, that is to say, under a certain class of morphisms, as well as properties which are stated relative to a particular commodity space.

Let C be a set of commodity labels and let

$$F_C=\{(m,n)\mid m,n \text{ are non-empty, disjoint, finite subsets of } C\}$$

Further, let $\Sigma(C)$ be the set of all non-empty, closed, convex subsets A, of $R^m xR^n$, for $(m,n)\in F_C$, such that A has 0 as a boundary point and $R^m_+ xR^n_+\subset A$.

A <u>family of approximations</u> on $\Sigma(C)$ is a mapping Z assigning to each triple $(a;p^m,p^n)\in\bar{R}_+ xR^m_{++}xR^n_{++}$ an element $Z(a;p^m,p^n)$ of $\Sigma(C)$ having the property that

$$(p^m,p^n)(z^m,z^n)\geq 0 \text{ for } (z^m,z^n)\in Z(a;p^m,p^n)$$

Thus each set in the approximating family is closed, convex with 0 as a boundary point. It is supported by the corresponding prices and a may be interpreted as a measure of curvature; larger values corresponding to stronger curvature.

Consider now two sets of commodity labels C and C'. Presumably there should be some relation between an approximating family on R^C and on $R^{C'}$.

Let (η,α) be a market morphism with $\gamma:C \to C'$ a bijection and $\alpha = 1_C$. Then, for $(m,n)\varepsilon F_C$ we have $(\gamma(m),\gamma(n))\varepsilon F_{C'}$, and $|\gamma(m)|=|m|$, $|\gamma(n)|=|n|$. We then demand that

$$Z(a;\tilde{p}^{\gamma(m)},\tilde{p}^{\gamma(n)})=g_{\eta,\alpha}(Z(a;p^m,p^n))$$

where $\tilde{p}_{c'}=p_{\gamma^{-1}(c')}$ for $c'\varepsilon C'$.

This condition implies, firstly, that the family of approximations on R^C is related to the family of approximations on $R^{C'}$ in a reasonable way. Secondly, choosing C=C', the condition is seen to imply a commodity label invariance; it is irrelevant where in C the sets m and n are located. Also , choosing C=C' and γ so that $\gamma(m)=m$ and $\gamma(n)=n$ the condition is seen to imply a symmetry property or, equivalently, $Z(a;p^m,p^n)$ is independent of the labels chosen for the goods in m and n.

If, as sems natural, we restrict ourselves to families of approximations satisfying the invariance property stated above then, there is no loss of generality in considering a fixed commodity space.

The invariance property implies that

$$\{Z(a;1_m,1_n) \mid |m|=1,\ |n|=1 \text{ and } a\varepsilon\dot{R}^m\}$$

may be regarded as a family of sets in R^2. This family, which will eventually turn out to be the family studied in Chapter 5, will be called the <u>corresponding family</u> and will be denoted by $\tilde{Z}$.

Next, consider the following axiom.

AXIOM V. $Z(a;1_m,1_n)=\{(z^m,z^n)eR^m xR^n \mid (1_m z^m,1_n z^n)\varepsilon\tilde{Z}(1;1,1)\}$

(Here 1_m (1_n) is the price vector with coordinates 1 for all $c\varepsilon m$ ($c\varepsilon n$)) One justification for Axiom V runs as follows. Consider a consumer having $Z(a;p^m,p^n)$ as his set of improvements and suppose that he is unable to distinguish between each of the goods in m and between each of the goods in n. Note that neither does the market distinguish between goods since they have identical prices.

A morphism (η,α) where $\alpha=1_C$, $\gamma:C\to C'$ such that $\gamma(m)=\{c_1'\}$ and $\gamma(n)=\{c_2'\}$ would take the set of improvements to

$$\{(z^{c_1'},z^{c_2'})\,|\,z^{c_1'}=1_m z^m,\ z^{c_2'}=1_n z^n\}\subset R^{\{c_1'\}}xR^{\{c_2'\}}$$

Axiom V therefore amounts to the assumption that the measure of curvature describing substitution in improvement possibilities should be independent of the chosen description of the consumer.

To give a rationale for the next axiom, which is just a many-goods version of Axiom III in Chapter 5, consider a morphism (η,α) with $\gamma=id_C$. As noted in Example 9.1, the mapping $g_{\eta,\alpha}$ may then be interpreted as a change in the units of measurement. The set $Z(a;1_m,1_n)$ is mapped by $g_{\eta,\alpha}$ to

$$\{(\tilde{z}^m,\tilde{z}^n)\,|\,\exists(z^m,z^n)\varepsilon Z(a;1_m,1_n)\text{ such that }\tilde{z}_c^m=\alpha_c z^m,\ \tilde{z}_c^n=\alpha_c z^n\text{ for }c\varepsilon m\cup n\}$$

Axiom III' asserts that such a change in the units of measurement should not affect the measure of curvature, a.

AXIOM III'. For $\alpha\varepsilon R^C_{++}$, $(a;p^m,p^n)\varepsilon\bar{R}_+xR^m_{++}xR^n_{++}$ and $(m,n)\varepsilon F_C$

$$Z(a;(\alpha_c p_c^m)_{c\varepsilon m},(\alpha_c p_c^n)_{c\varepsilon n})=$$

$$=\{(z^m,z^n)\varepsilon R^{m\cup n}\,|\,\exists(x^m,x^n)\varepsilon Z(a;p^m,p^n)\text{ and }z_c^m=\frac{x^m}{\alpha_c},\ z_c^n=\frac{x^n}{\alpha_c}\}$$

We now have

THEOREM 9.4. Let Z be a mapping satisfying Axiom III' and V such that the corresponding family $\tilde{Z}$ satisfies I-IV of Section 5.1. Then

$$Z(a;p^m,p^n)=\{(z^m,z^n)\varepsilon R^m xR^n\,|\,p^m z^m+p^n z^n+a(p^m z^m)(p^n z^n)\geq 0\text{ and }p^m z^m+p^n z^n\geq 0\}$$

for $a\varepsilon R_+$, $p^m\varepsilon R^m_{++}$, $p^n\varepsilon R^n_{++}$, $(m,n)\varepsilon F_C$ and

$$Z(\infty;p^m,p^n)=\{(z^m,z^n)\varepsilon R^m xR^n\,|\,(p^m z^m,p^n z^n)\varepsilon R^2_+\}.$$

Proof: Theorem 5.1 gives that

$$Z(a;1,1)=\{(x,y)|x+y+axy\geq 0,\ x+y\geq 0\}$$

Applying Axiom III' and V we have, for $a\varepsilon R_+$, $p^m \varepsilon R^m_{++}$, $p^n \varepsilon R^n_{++}$

$$Z(a;p^m,p^n)=\{(z^m,z^n)\varepsilon R^m xR^n|x^m=p^m z^m,\ x^n=p^n z^n \text{ and } (x^m,x^n)\varepsilon Z(a;p^m,p^n)\}=$$

$$=\{(z^m,z^n)\varepsilon R^m xR^n|z^m=\frac{x^m_c}{p^m_c},\ z^n=\frac{x^n_c}{p^n_c} \text{ and } (1_m x^m,1_n x^n)\varepsilon Z(a;1,1)\}=$$

$$=\{(z^m,z^n)\varepsilon R^m xR^n|p^m z^m+p^n z^n+a(p^m z^m)p^n z^n)\geq 0,\ p^m z^m+p^n z^n\geq 0\}$$

The case $a=\infty$ is left to the reader. ●

Note that the composition of sets, as defined in Section 8.2, translates to addition of parameters, that is,

$$Z(a;p^m,p^n)oZ(b;p^n,p^k)=Z(a+b;p^m,p^k)$$

for $(m,n),(n,k)$ and $(m,k)\varepsilon F_C$.

For the subfamily $\tilde{Z}$ this is, of course, Axiom I. For arbitrary sets of the approximating family defined in Theorem 9.4 it follows from the relation

$$Z(a;p^m,p^n)oZ(b;p^n,p^k)=\{(z^m,z^k)\varepsilon R^{m\cup k}|\exists z^n\varepsilon R^n \text{ so that}$$

$$p^m z^m+p^n z^n+a(p^m z^m)(p^n z^n)\geq 0\ ,\ p^m z^m+p^n z^n\geq 0 \text{ and}$$

$$p^n z^n+p^k z^k+b(p^n z^n)(p^k z^k)\geq 0\ ,\ p^n z^n+p^k z^k\geq\}=$$

$$\{(z^m,z^k)\varepsilon R^{m\cup k}|p^m z^m+p^k z^k+(a+b)(p^m z^m)(p^k z^k)\geq 0\}$$

where the first equality follows from the definition of o and the last from the considerations in Example 3.7.

Axiom IV is applicable only to the family $\tilde{Z}$ but it is easy to see that a generalized version of Axiom II holds for Z.

Given the family of sets, $Z(a;p^m,p^n)$, we may define the <u>inner and outer curvature</u> of a set $A\subset R^m xR^n$, $(m,n)\varepsilon F_C$, supported by (p^m,p^n) as

$$m_{(p^m,p^n)}(A)=\inf\{a|Z(a;p^m,p^n)\subset A\}$$

$$M_{(p^m,p^n)}(A)=\sup\{a|A\subset Z(a;p^m,p^n)\}$$

Obviously the concepts so defined depend not only on the set A, but also on the partition of $m \cup n$ into "first" and "last" subsets, m and n. In the applications, however, this partition will always arise in a natural way.

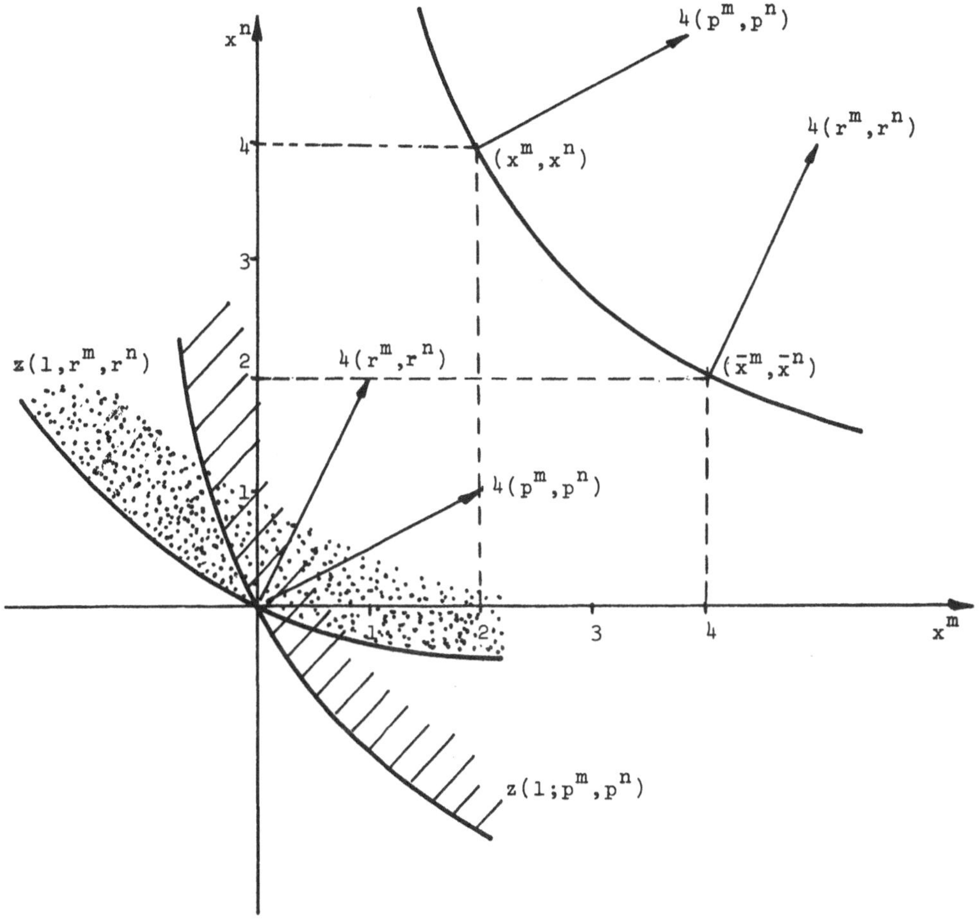

FIGURE 9.1

In the one-good case, Section 4.5, we showed that the family of approximations was generated by a utility function. Below we state a similar characterization result for the many-goods case. This characterization clearly reveals the two-dimensional character of the measure of curvature which, by the way, admits a geometrical interpretation. In Figure 9.1 we have shown for the case $|m|=|n|=1$, $p^m=1/2$, $p^n=1/4$ the sets ocurring in the statement of Theorem 9.5.(p^m and p^n have been multiplied by 4 in the figure.)

THEOREM 9.5. Consider the family of utility functions

$$U(p^m,p^n,\cdot,:):R^m x R^n \longrightarrow R$$

where $(m,n)\varepsilon F_C$, $p^m \varepsilon R^m_{++}$, $p^n \varepsilon R^n_{++}$ and

$$U(p^m,p^n,x^m,x^n)=(p^m x^m)(p^n x^n)$$

Let $(\tilde{x}^n,\tilde{x}^m)$ be a point such that $p^m\tilde{x}^m=p^n\tilde{x}^n=1$ and $(\bar{x}^m,\bar{x}^n)$ any point such that $(p^m\bar{x}^m)(p^n\bar{x}^n)=1$

Put

$$\frac{\partial U}{\partial x^m_c}(x^m,x^n)=p^m_c(p^n x^n)=r^m_c \quad \text{for } c\varepsilon m$$

$$\frac{\partial U}{\partial x^n_c}(x^m,x^n)=(p^m x^m)p^n_c=r^n_c \quad \text{for } c\varepsilon n$$

Then

$$Z(1;r^m,r^n)=\{(x^m,x^n)\varepsilon R^m x R^n \mid U(p^m,p^n,x^m,x^n)\geq 1\}-(\bar{x}^m,\bar{x}^n)$$

and

$$Z(1;p^m,p^n)+(\tilde{x}^m,\tilde{x}^n)=Z(1;r^m,r^n)+(\bar{x}^m,\bar{x}^n)$$

9.3. APPLICATIONS OF THE MEASURE OF CURVATURE

We turn now to some applications of approximating families and , through this, the measure of curvature. Counterparts of the results in Chapter 4 will be derived. The essentially two-dimensional character of the curvature measure permits us to draw heavily on the results of Chapter 5.

We have seen that efficiency considerations amount essentially to the investigation of substitution possibilitites between disjoint subsets of the commodity space. When the sets describing the substitution possibilities belong to the family Z two cases of interest can occur. This is described by the following lemma which, in its turn, is later on used to derive two efficiency criteria by approximation.

LEMMA 9.6. Let C be a commodity space and let $(m(i),n(i))\varepsilon F_C$, $d_i \varepsilon R_+$ for $i\varepsilon N$. Consider the sequence of sets

$$Z(d_i;p^{m(i)},p^{n(i)})\subset R^{m(i)\cup n(i)} \quad , \ i\varepsilon N$$

(i) If $d_i \longrightarrow \infty$ as $i \longrightarrow \infty$ then for any sequence

$$(z^{m(i)},z^{n(i)})\varepsilon Z(d_i;p^{m(i)},p^{n(i)}), \ i\varepsilon N$$

we have

$$\liminf p^{m(i)}z^{m(i)} \geq 0$$

(ii) If $(d_i)_{i\varepsilon N}$ is bounded from above, then there exists a sequence

$$(z^{m(i)},z^{n(i)})\varepsilon Z(a;p^{m(i)},p^{n(i)})$$

such that $z^{m(i)}\varepsilon R_{++}^{m(i)}$ for $i\varepsilon N$ and

$$p^{m(i)}z^{m(i)} \leq -\delta < 0 \text{ for some } \delta > 0.$$

<u>Proof</u>: We have, for $d_i > 0$,

$$\inf\{p^{m(i)}z^{m(i)} \mid p^{m(i)}z^{m(i)}+p^{n(i)}z^{n(i)}+a(p^{m(i)}z^{m(i)})(p^{n(i)}z^{n(i)}) \geq 0 \text{ and}$$

$$p^{m(i)}z^{m(i)}+p^{n(i)}z^{n(i)} \geq 0\} = -\frac{1}{d_i}. \ \bullet$$

THEOREM 9.7. Let $\Sigma=(S^m)_{m\varepsilon M}$ be a reduced model with support $(p_c)_{c\varepsilon C}$ and let M(m') be the m'-canonical market structure, $M(m')=\{m',m_1,m_2,...\}$.

For $i,T\varepsilon N\cup\{0\}$, $i\leq T$ let

$m_{i,T}$ denote the outer cuvature of $S^{m_i}o...oS^{m_T}$ and let

$m_{i,T}$ denote the inner cuvature of $S^{m_i}o...oS^{m_T}$

(i) If $\inf_{i\in N} \sup_{T\geq i} m_{i,T}=\infty$ then Σ is efficient.

(ii) If $\inf_{i\in N} \sup_{T\geq i} M_{i,T}<\infty$ then Σ is inefficient.

Proof: Lemma 9.6 shows that we may construct an improving sequence for Σ if

$$\inf_{i\in N} \sup_{T\geq i} M_{i,T}<\infty$$

and that there can be no improving sequence if

$$\inf_{i\in N} \sup_{T\geq i} m_{i,T}=\infty. \bullet$$

COROLLARY. Let (C,M) be a canonical market structure and $\Sigma=(S^m)_{m\in M}$ a reduced model with support $(p_c)_{c\in C}$. For $i\in N$ let

m_h denote the outer curvature of S^{m_h} and let

M_h denote the inner curvature of S^{m_h}

Assume that there is a $K\geq 0$ such that $M_h\leq Km_h$ for $h\in N$. Then Σ is efficient if and only if $\Sigma_{t=i}^{\infty} m_i$ is divergent for each $i\in N$

Proof: Defining $m_{i,T}$ and $M_{i,T}$ as in Theorem 9.7 we have, since the family of approximations satisfy the many-goods version of Axiom I,

$$m_{i,T}=m_i+m_{i+1}+...+m_T$$

$$M_{i,T}=M_i+M_{i+1}+...+M_T$$

If for some $i\in N$, $\Sigma_{t=i}^{\infty} m_i$ is finite, then since $M_h\leq Km_h$ for $h\in N$ we get

$$\inf_{i\in N} \sup_{T\geq i} m_{i,T}<\infty$$

and Σ is inefficient by Theorem 9.7(ii).

On the other hand, if $\Sigma_{t=i}^{\infty} m_h$ is divergent for each $i \in N$ then

$$\inf_{i \in N} \sup_{T \geq i} m_{i,T} = \infty$$

and Σ is efficient by Theorem 9.7. ●

Note that the crucial difference between the results for general market structures and generation market structures is that for general models we cannot relate the measure of curvature to the original sets of the model but only to the sets of some induced canonical market structure, unless of course the original model already has a canonical market structure.

9.4. VALUE MODELS

The preceding sections indicate that we may decide upon the efficiency/inefficiency for certain reduced models by considering suitably chosen one-good models. In this section we will show that to any reduced model defined on a canonical market structure there is a corresponding one-good model, which we shall call a value model, such that the efficiency of the value model implies the efficiency of the original model.

Let (C,M) be a canonical market structure with $M=\{m_0,m_1,m_2,...\}$ and $(S^m)_{m\in M}$ a reduced model with support $(p_c)_{c\in C}$. In Example 9.3 we have indicated how to construct a one-good model from $(S^m)_{m\in M}$ when (C,M) satisfied an additional assumption. As will be seen below this construction, slightly modified, may be used also in the general case.

First of all, define the reduced model $(\tilde{S}^m)_{m\in M}$ by

$$\tilde{S}^{m_i} = \{x \in S^{m_i} \mid x_c = 0 \text{ for } c \notin (m_i \cap m_{i+1}) \cup (m_i \cap m_{i-1})\}$$

Then $(\tilde{S}^m)_{m\in M}$ is obviously a reduced model which is efficient if and only if $(S^m)_{m\in M}$ is efficient. We may identify $(\tilde{S}^m)_{m\in M}$ with a reduced model on (C',M') where C' and M' are derived form C and M simply by deleting those (internal) goods in C which belong to only one market, that is,

$$m_i' = \{c \mid c \in (m_i \cap m_{i+1}) \cup (m_i \cap m_{i-1})\}$$

and

$$C' = \cup_{m_i' \in M'} m_i$$

The construction of Example 9.3 can now be applied to $(\tilde{S}^m)_{m \in M}$, or equivalently, we can define the one-good reduced model $Q(\Sigma) = (S_t)_{t \in N}$ corresponding to $\Sigma = (S^m)_{m \in M}$ by

$$S_t = \{(z_1, z_2) \mid \exists z \in S^{m_t} \text{ such that } z_c = 0 \text{ for } c \notin (m_t \cap m_{t+1}) \cup (m_t \cap m_{t-1})$$
$$\text{and } (z_1, z_2) = p^{m_t \cap m_{t-1}} z, p^{m_t \cap m_{t+1}} z)\}$$

$Q(\Sigma)$ is the <u>value model</u> corresponding to Σ.

The following result is immediate.

THEOREM 9.8. Let $\Sigma = (S^m)_{m \in M}$ be a reduced model over a market structure with support $(p_c)_{c \in C}$ and let $Q(\Sigma)$ be the corresponding value model. If $Q(\Sigma)$ is efficient then Σ is efficient.

<u>Proof</u>: Any improving sequence for Σ induces an improving sequence for $Q(\Sigma)$. ●

As an application we obtain a "Malinvaud Criterion" for reduced many-goods models.

COROLLARY. Under the assumptions of in Theorem 9.8, if

$$\liminf_{i \to \infty} [\inf \{p^{m_i \cap m_{i-1}} z \mid z \in S^{m_i}\}] = 0$$

then Σ is efficient.

<u>Proof</u>: Theorem 9.8 and Theorem 4.5. ●

The fact that we cannot without further assumptions conclude from inefficiency of the value model to inefficiency of the original model, is shown by the following example.

EXAMPLE 9.9. Let C=N and

$m_o=\{1,2\}$
$m_1=\{1,2,3,4\}$
$m_2=\{3,4,5,6\}$
.
.
$m_t=\{2t-1,2t,2t+1,2t+2\}$ for $t\geq 1$

$(S^m)_{m\varepsilon M}$ is defined by

$$S^{m_o}=R_+^{m_o}$$

$$S^{m_t}=\{(x_{2t-1},x_{2t},x_{2t+1},x_{2t+2})\,|\,x_{2t-1}\geq 0,\ x_{2t+1}\geq 0 \text{ and } x_{2t}+x_{2t+1}\geq 0\}$$

for t odd and, for t even,

$$S^{m_t}=\{(x_{2t-1},x_{2t},x_{2t+1},x_{2t+2})\,|\,x_{2t}\geq 0,\ x_{2t+2}\geq 0 \text{ and } x_{2t-1}+x_{2t+1}\geq 0\}$$

Then $p_c=1$ for $c\varepsilon C$ defines a support to $(S^m)_{m\varepsilon M}$. The corresponding value model is

$S_o=R_+$ and $S_t=\{(x,y)\,|\,x+y\geq 0\}$ for $t\varepsilon N$

Hence $(S_t)_{t\varepsilon N}$ is inefficient. On the other hand, it is easy to see that there can be no improving sequence for $(S^m)_{m\varepsilon M}$ which is thus efficient.●

The considerations above, which show that value models reflect efficiency properties of the original model but that inefficiency of the value model is in general no guarantee for the inefficiency of the original model, lead to the natural problem of characterizing the class of reduced models for which the value model is sufficient information in deciding upon the efficiency/inefficiency question.

We shall not go into a detailed discussion of this question but only point out that essentially it is another formulation of the condition which we have already met as Axiom V, the invariance property under market morphisms of the sets S^m making up the model under consideration. Thus we see how the approximating family Z (and implictly the hyperbolic measure of curvature) comes up repeatedly when we attack our problem from different angles.

REFERENCES

Arrow, K. and Hahn, F. [1971] General Competitive Analysis, Holden-Day, San Francisco.

Balasko, Y. and Shell, K. [1980] , The Overlapping-Generations model, I: the Case of Pure Exchange without Money, Journal of Economic Theory 23, 281-306.

Balasko, Y. and Shell, K. [1981,a] , The Overlapping-Generations Model, II: The Case of Pure Exchange with Money, Journal of Economic Theory 24, 112-142.

Balasko, Y. and Shell, K. [1981,b] , The Overlapping-Generations Model, III: The Case of Log-Linear Utility Functions, Journal of Economic Theory 24, 143-152.

Balasko, Y., Cass, D. and Shell, K. [1980], Existence of Competitive Equilibrium in a General Overlapping Generations Model, Journal of Economic Theory 23, 307-322.

Benveniste, L.M. [1976,a] , Two notes on the Malinvaud Condition for Efficiency of Infinite Horizon Programs, Journal of Economic Theory 12, 338-346.

Benveniste, L.M. [1976,b] , A Complete Characterization of Efficiency for a General Capital Accumulation Model, Journal of Economic Theory 12, 325-337.

Benveniste, L.M. and Gale, D. [1975] , An Extension of Cass' Characterization of Infinite Efficient Production Programs, Journal of Economic Theory 10, 229-238.

Benveniste, L.M. and Mitra, T. [1979] , Characterizing Inefficiency of Infinite Horizon Programs in Nonsmooth Technologies, in: Green, J.R. and Scheinkman, J.A.(Eds.), General Equilibrium, Growth and Trade, Academic Press Inc.

Bergstrom, T.C. [1976] , How to Discard "Free Disposability" at No Cost, Journal of Mathematical Economics 3, 131-134

Borglin, A. [1975] , The Theory of Lindahl Equilibria Derived from the Theory of Uncertainty, Meddelande 1975:5,Nationalekonomiska Institutionen,Lund

Borglin, A. and Keiding, H. [1981] , Existence of Equilibrium without Walras' Law, in Moeshlin, O. and Pallaschke, D. (Eds.); Game Theory and Mathematical Economics, North-Holland 1981.

Borglin, A. and Keiding, H. [1983] , Efficiency in One-Sector, Discrete Time, Infinite Horizon Models, Journal of Economic Theory 33, 183-196

Cass, D. [1972,a] , On Capital Overaccumulation in the Aggregative Neoclassical Model of Economic Growth: A Complete Characterization, Journal of Economic Theory 4, 200-223.

Cass, D. [1972,b] , Distinguishing Inefficient Competitive Growth Paths: A Note on Capital Overaccumulation and Rapidly Diminishing Future Values of Consumption in a Fairly General Model of Capitalistic Production, Journal of Economic Theory 4, 224-240.

Cass, D. and Yaari, M.E. [1966] , A Re-examination of the Pure Consumption Loans Model, Journal of Political Economy 74, 353-367.

Cass, D., Okuno, M. and Zilcha, I. [1979] , The Role of Money in Supporting Pareto Optimality of Competitive Equilibrium in Consumption-Loan Type Models, Journal of Economic Theory 20, 41-80.

Chang, W.W., Kemp, M.C. and Van Long, N. [1983] , Dynamic Properties of a Simple Overlapping-Generations Model, Oxford Economic Papers 35, 366-373.

Clark, S.A. [1979] , Pareto Optimality in the Pure Distribution Economy with an Infinite Number of Consumers and Commodities, Journal of Economic Theory 21, 336-347.

Clark, S.A. [1981] , A Combinatorial Analysis of the Overlapping Generations Model, Review of Economic Studies 48, 139-145.

Debreu, G. [1962] ., New Techniques in Equilibrium Analysis, International Economic Review 3,257-274

Debreu, G. [1959] , Theory of Value, Wiley, N.Y.

Diamond, P.A. [1965] , National Debt in a Neo-Classical Growth Model, American Economic Review 55, 1126-1150.

Feller, W. [1968] An Introduction to Probability Theory and Its Applications, vol.1, 3rd edition, Wiley 1968

Grandmont, M. and Laroque, G. [1973] , Money in the Pure Consumption Loan Model, Journal of Economic Theory 6, 382-395.

Hildenbrand, W. [1974] , Core and Equilibria in a Large Economy, Princeton Un. Press 1974

Koopmans, T. [1957] , Three Essays on the State of Economic Science, McGraw-Hill, N.Y.

Kurz, M. and Starrett, D. [1970] , On the Efficiency of Competitive Programmes in an Infinite Horizon Model, Review of Economic Studies 37, 571-584.

Laugwitz, D. [1965] , Differential and Riemannian Geometry, Academic Press 1965.

Majumdar, M., Mitra, T. and McFadden, D. [1976] , On Efficiency and Pareto Optimality of Competitive Programs in Closed Multisector Models, Journal of Economic Theory 13, 26-46.

Malinvaud, E. [1953] , Capital Accumulation and Efficient Allocation of Resources, Econometrica 21, 233-268.

Malinvaud, E. [1962] , Efficient Capital Accumulation: A Corrigendum, Econometrica 30, 570-573.

McFadden, D. [1967] , The Evaluation of Development Programs, Review of Economic Studies 34, 25-50.

McFadden, D., Mitra, T and Majumdar, M. [1980] , Pareto Optimality and Competitive Equilbrium in Infinite Horizon Economies, Journal of Mathematical Economics 7, 1-26.

Mitra, T. [1976] , On Efficient Capital Accummulation in a Multisector Neoclassical Model, Review of Economic Studies 43, 423-429.

Mitra, T. [1979,a] , On the Value Maximizing Property of Infinite Horizon Efficient Programs, International Economic Review 20, 635-642.

Mitra, T. [1979,b] , Identifying Inefficiency in Smooth Aggregative Models of Economic Growth: A Unifying Criterion, Journal of Mathematical Economics, 85-111.

Mitra, T. [1981] , Efficiency, Weak Value Maximality and Weak Value Optimality in a Multisector Model, Review of Economic Studies 48, 643-647.

Mitra, T. and Majumdar, M. [1976] , A Note on the Role of the Transversality Condition in Signalling Capital Overaccumulation, Journal of Economic Theory 13, 47-57.

Okuno, M. and Zilcha, I. [1980] , On the Efficiency of a Competitive Equilibrium in Infinite Horizon Monetary Economies, Review of Economic Studies 47, 797-807.

Peleg, B. [1970] , Efficiency Prices for Optimal Consumption Plans, Journal of Mathematical Analysis and Applications 29, 83-90.

Peleg, B. [1971] , Efficiency Prices for Optimal consumption Plans II, Israel Journal of Mathematics 9, 222-234.

Peleg, B. and Yaari, M.E. [1970] , Efficiency Prices in an Infinite-Dimensional Space, Journal of Economic Theory 2, 41-85.

Phelps, E.S. [1965] , Second Essay on the Golden Rule of Accumulation, American Economic Review 55, 793-814.

Phelps, E.S. [1966] , Golden Rules of Economic Growth: An Essay on Dynamical Efficiency, Norton, N.Y.

Samuelson, P.A. [1958] , An Exact Consumption-Loan Model of Interest with or without the Social Contrivance of Money, Journal of Political Economy 66, 467-482.

Samuelson, P.A. [1959] , Reply, Journal of Political Economy 67, 518-522.

Samuelson, P.A. [1960] , Infinity, Unanimity and Singularity: A Reply, Journal of Political Economy 68, 76-82.

Shell, K. [1971] , Notes on the Economics of Infinity, Journal of Political Economy 79, 1002-1011.

Starrett, D.A. [1970] , On Some Efficiency Characteristics of a General Production Model, International Economic Review 11, 506-520.

Starrett, D.A. [1972] , On Golden Rules, the "Biological Theory of Interest" and Competitive Inefficiency, Journal of Political Economy 80, 276-291.

Swan, T.W. [1956] , Economic Growth and Capital Accumulation, Economic Record 32, 343-361

Weitzmann, M. [1973] Duality Theory for Infinite Horizon Models, Management Science 19, 783-789

Wilson, C.A. [1981] , Equilibrium in Dynamic Models with an Infinity of Agents, Journal of Economic Theory 24, 95-111.

Von Weizsäcker, C.C. [1965] , Existence of Optimal Programs of Accumulation for an Infinite Time Horizon, Review of Economic Studies 32, 85-104

INDEX

Vol. 184: R. E. Burkard and U. Derigs, Assignment and Matching Problems: Solution Methods with FORTRAN-Programs. VIII, 148 pages. 1980.

Vol. 185: C. C. von Weizsäcker, Barriers to Entry. VI, 220 pages. 1980.

Vol. 186: Ch.-L. Hwang and K. Yoon, Multiple Attribute Decision Making – Methods and Applications. A State-of-the-Art-Survey. XI, 259 pages. 1981.

Vol. 187: W. Hock, K. Schittkowski, Test Examples for Nonlinear Programming Codes. V. 178 pages. 1981.

Vol. 188: D. Bös, Economic Theory of Public Enterprise. VII, 142 pages. 1981.

Vol. 189: A. P. Lüthi, Messung wirtschaftlicher Ungleichheit. IX, 287 pages. 1981.

Vol. 190: J. N. Morse, Organizations: Multiple Agents with Multiple Criteria. Proceedings, 1980. VI, 509 pages. 1981.

Vol. 191: H. R. Sneessens, Theory and Estimation of Macroeconomic Rationing Models. VII, 138 pages. 1981.

Vol. 192: H. J. Bierens: Robust Methods and Asymptotic Theory in Nonlinear Econometrics. IX, 198 pages. 1981.

Vol. 193: J. K. Sengupta, Optimal Decisions under Uncertainty. VII, 156 pages. 1981.

Vol. 194: R. W. Shephard, Cost and Production Functions. XI, 104 pages. 1981.

Vol. 195: H. W. Ursprung, Die elementare Katastrophentheorie. Eine Darstellung aus der Sicht der Ökonomie. VII, 332 pages. 1982.

Vol. 196: M. Nermuth, Information Structures in Economics. VIII, 236 pages. 1982.

Vol. 197: Integer Programming and Related Areas. A Classified Bibliography. 1978 – 1981. Edited by R. von Randow. XIV, 338 pages. 1982.

Vol. 198: P. Zweifel, Ein ökonomisches Modell des Arztverhaltens. XIX, 392 Seiten. 1982.

Vol. 199: Evaluating Mathematical Programming Techniques. Proceedings, 1981. Edited by J.M. Mulvey. XI, 379 pages. 1982.

Vol. 200: The Resource Sector in an Open Economy. Edited by H. Siebert. IX, 161 pages. 1984.

Vol. 201: P. M. C. de Boer, Price Effects in Input-Output-Relations: A Theoretical and Empirical Study for the Netherlands 1949–1967. X, 140 pages. 1982.

Vol. 202: U. Witt, J. Perske, SMS – A Program Package for Simulation and Gaming of Stochastic Market Processes and Learning Behavior. VII, 266 pages. 1982.

Vol. 203: Compilation of Input-Output Tables. Proceedings, 1981. Edited by J. V. Skolka. VII, 307 pages. 1982.

Vol. 204: K. C. Mosler, Entscheidungsregeln bei Risiko: Multivariate stochastische Dominanz. VII, 172 Seiten. 1982.

Vol. 205: R. Ramanathan, Introduction to the Theory of Economic Growth. IX, 347 pages. 1982.

Vol. 206: M. H. Karwan, V. Lotfi, J. Telgen, and S. Zionts, Redundancy in Mathematical Programming. VII, 286 pages. 1983.

Vol. 207: Y. Fujimori, Modern Analysis of Value Theory. X, 165 pages. 1982.

Vol. 208: Econometric Decision Models. Proceedings, 1981. Edited by J. Gruber. VI, 364 pages. 1983.

Vol. 209: Essays and Surveys on Multiple Criteria Decision Making. Proceedings, 1982. Edited by P. Hansen. VII, 441 pages. 1983.

Vol. 210: Technology, Organization and Economic Structure. Edited by R. Sato and M. J. Beckmann. VIII, 195 pages. 1983.

Vol. 211: P. van den Heuvel, The Stability of a Macroeconomic System with Quantity Constraints. VII, 169 pages. 1983.

Vol. 212: R. Sato and T. Nôno, Invariance Principles and the Structure of Technology. V, 94 pages. 1983.

Vol. 213: Aspiration Levels in Bargaining and Economic Decision Making. Proceedings, 1982. Edited by R. Tietz. VIII, 406 pages. 1983.

Vol. 214: M. Faber, H. Niemes und G. Stephan, Entropie, Umweltschutz und Rohstoffverbrauch. IX, 181 Seiten. 1983.

Vol. 215: Semi-Infinite Programming and Applications. Proceedings, 1981. Edited by A. V. Fiacco and K. O. Kortanek. XI, 322 pages. 1983.

Vol. 216: H. H. Müller, Fiscal Policies in a General Equilibrium Model with Persistent Unemployment. VI, 92 pages. 1983.

Vol. 217: Ch. Grootaert, The Relation Between Final Demand and Income Distribution. XIV, 105 pages. 1983.

Vol. 218: P. van Loon, A Dynamic Theory of the Firm: Production, Finance and Investment. VII, 191 pages. 1983.

Vol. 219: E. van Damme, Refinements of the Nash Equilibrium Concept. VI, 151 pages. 1983.

Vol. 220: M. Aoki, Notes on Economic Time Series Analysis: System Theoretic Perspectives. IX, 249 pages. 1983.

Vol. 221: S. Nakamura, An Inter-Industry Translog Model of Prices and Technical Change for the West German Economy. XIV, 290 pages. 1984.

Vol. 222: P. Meier, Energy Systems Analysis for Developing Countries. VI, 344 pages. 1984.

Vol. 223: W. Trockel, Market Demand. VIII, 205 pages. 1984.

Vol. 224: M. Kiy, Ein disaggregiertes Prognosesystem für die Bundesrepublik Deutschland. XVIII, 276 Seiten. 1984.

Vol. 225: T. R. von Ungern-Sternberg, Zur Analyse von Märkten mit unvollständiger Nachfragerinformation. IX, 125 Seiten. 1984

Vol. 226: Selected Topics in Operations Research and Mathematical Economics. Proceedings, 1983. Edited by G. Hammer and D. Pallaschke. IX, 478 pages. 1984.

Vol. 227: Risk and Capital. Proceedings, 1983. Edited by G. Bamberg and K. Spremann. VII, 306 pages. 1984.

Vol. 228: Nonlinear Models of Fluctuating Growth. Proceedings, 1983. Edited by R. M. Goodwin, M. Krüger and A. Vercelli. XVII, 277 pages. 1984.

Vol. 229: Interactive Decision Analysis. Proceedings, 1983. Edited by M. Grauer and A. P. Wierzbicki. VIII, 269 pages. 1984.

Vol. 230: Macro-Economic Planning with Conflicting Goals. Proceedings, 1982. Edited by M. Despontin, P. Nijkamp and J. Spronk. VI, 297 pages. 1984.

Vol. 231: G. F. Newell, The M/M/∞ Service System with Ranked Servers in Heavy Traffic. XI, 126 pages. 1984.

Vol. 232: L. Bauwens, Bayesian Full Information Analysis of Simultaneous Equation Models Using Integration by Monte Carlo. VI, 114 pages. 1984.

Vol. 233: G. Wagenhals, The World Copper Market. XI, 190 pages. 1984.

Vol. 234: B. C. Eaves, A Course in Triangulations for Solving Equations with Deformations. III, 302 pages. 1984.

Vol. 235: Stochastic Models in Reliability Theory. Proceedings, 1984. Edited by S. Osaki and Y. Hatoyama. VII, 212 pages. 1984.

Vol. 236: G. Gandolfo, P. C. Padoan, A Disequilibrium Model of Real and Financial Accumulation in an Open Economy. VI, 172 pages. 1984.

Vol. 237: Misspecification Analysis. Proceedings, 1983. Edited by T. K. Dijkstra. V, 129 pages. 1984.

Vol. 238: W. Domschke, A. Drexl, Location and Layout Planning. IV, 134 pages. 1985.

Vol. 239: Microeconomic Models of Housing Markets. Edited by K. Stahl. VII, 197 pages. 1985.

Vol. 240: Contributions to Operations Research. Proceedings, 1984. Edited by K. Neumann and D. Pallaschke. V, 190 pages. 1985.

Vol. 241: U. Wittmann, Das Konzept rationaler Preiserwartungen. XI, 310 Seiten. 1985.

Vol. 242: Decision Making with Multiple Objectives. Proceedings, 1984. Edited by Y. Y. Haimes and V. Chankong. XI, 571 pages. 1985.

Vol. 243: Integer Programming and Related Areas. A Classified Bibliography 1981–1984. Edited by R. von Randow. XX, 386 pages. 1985.

Vol. 244: Advances in Equilibrium Theory. Proceedings, 1984. Edited by C. D. Aliprantis, O. Burkinshaw and N. J. Rothman. II, 235 pages. 1985.

Vol. 245: J. E. M. Wilhelm, Arbitrage Theory. VII, 114 pages. 1985.

Vol. 246: P. W. Otter, Dynamic Feature Space Modelling, Filtering and Self-Tuning Control of Stochastic Systems. XIV, 177 pages. 1985.

Vol. 247: Optimization and Discrete Choice in Urban Systems. Proceedings, 1983. Edited by B. G. Hutchinson, P. Nijkamp and M. Batty. VI, 371 pages. 1985.

Vol. 248: Plural Rationality and Interactive Decision Processes. Proceedings, 1984. Edited by M. Grauer, M. Thompson and A. P. Wierzbicki. VI, 354 pages. 1985.

Vol. 249: Spatial Price Equilibrium: Advances in Theory, Computation and Application. Proceedings, 1984. Edited by P. T. Harker. VII, 277 pages. 1985.

Vol. 250: M. Roubens, Ph. Vincke, Preference Modelling. VIII, 94 pages. 1985.

Vol. 251: Input-Output Modeling. Proceedings, 1984. Edited by A. Smyshlyaev. VI, 261 pages. 1985.

Vol. 252: A. Birolini, On the Use of Stochastic Processes in Modeling Reliability Problems. VI, 105 pages. 1985.

Vol. 253: C. Withagen, Economic Theory and International Trade in Natural Exhaustible Resources. VI, 172 pages. 1985.

Vol. 254: S. Müller, Arbitrage Pricing of Contingent Claims. VIII, 151 pages. 1985.

Vol. 255: Nondifferentiable Optimization: Motivations and Applications. Proceedings, 1984. Edited by V. F. Demyanov and D. Pallaschke. VI, 350 pages. 1985.

Vol. 256: Convexity and Duality in Optimization. Proceedings, 1984. Edited by J. Ponstein. V, 142 pages. 1985.

Vol. 257: Dynamics of Macrosystems. Proceedings, 1984. Edited by J.-P. Aubin, D. Saari and K. Sigmund. VI, 280 pages. 1985.

Vol. 258: H. Funke, Eine allgemeine Theorie der Polypol- und Oligopolpreisbildung. III, 237 pages. 1985.

Vol. 259: Infinite Programming. Proceedings, 1984. Edited by E. J. Anderson and A. B. Philpott. XIV, 244 pages. 1985.

Vol. 260: H.-J. Kruse, Degeneracy Graphs and the Neighbourhood Problem. VIII, 128 pages. 1986.

Vol. 261: Th. R. Gulledge, Jr., N. K. Womer, The Economics of Made-to-Order Production. VI, 134 pages. 1986.

Vol. 262: H. U. Buhl, A Neo-Classical Theory of Distribution and Wealth. V, 146 pages. 1986.

Vol. 263: M. Schäfer, Resource Extraction and Market Structure. XI, 154 pages. 1986.

Vol. 264: Models of Economic Dynamics. Proceedings, 1983. Edited by H. F. Sonnenschein. VII, 212 pages. 1986.

Vol. 265: Dynamic Games and Applications in Economics. Edited by T. Başar. IX, 288 pages. 1986.

Vol. 266: Multi-Stage Production Planning and Inventory Control. Edited by S. Axsäter, Ch. Schneeweiss and E. Silver. V, 264 pages. 1986.

Vol. 267: R. Bemelmans, The Capacity Aspect of Inventories. IX, 165 pages. 1986.

Vol. 268: V. Firchau, Information Evaluation in Capital Markets. VII, 103 pages. 1986.

Vol. 269: A. Borglin, H. Keiding, Optimality in Infinite Horizon Economies. VI, 180 pages. 1986.